I0505418

Contents

1 Greek alphabet and various symbols   **5**
  1.1 Greek alphabet 5
  1.2 Various symbols 6

2 Sets and Logic   **7**
  2.1 Sets 7
    2.1.1 Definition 7
    2.1.2 Union and Intersection 7
    2.1.3 Difference and Complement 8
    2.1.4 Inclusion and equality 8
    2.1.5 Properties 8
  2.2 Sets of numbers 9
    2.2.1 Intervals on $\mathbb{R}$ 9
  2.3 Cartesian Product of Two Sets 10
  2.4 Cardinality of Sets 10
    2.4.1 Countable and Uncountable Sets 11
    2.4.2 Cardinality of the Continuum 12
  2.5 Logic 13
    2.5.1 The liar paradox - Statements 13
    2.5.2 Truth tables and propositional connectives 13
    2.5.3 Statement Forms 14
    2.5.4 Tautology 15
    2.5.5 Quantifiers, notations 16
    2.5.6 Quantifiers, properties 16
    2.5.7 Direct proof 16
    2.5.8 Proof by contraposition 17
    2.5.9 Proof by contradiction 17
    2.5.10 Proof by equivalence 17
    2.5.11 Proof by mathematical induction 17
    2.5.12 Some conjectures 18

3 Group Theory   **19**
  3.1 Closed binary operation 19
    3.1.1 Some definitions 19
  3.2 Group axioms 20

Chapter 1

Greek alphabet and various symbols

1.1   Greek alphabet

Lowercase	Uppercase	Name
α	A	*alpha*
β	B	*bêta*
γ	Γ	*gamma*
δ	Δ	*delta*
ε or ε	E	*epsilon*
ζ	Z	*zêta*
η	H	*êta*
θ or ϑ	Θ	*thêta*
ι	I	*iota*
κ	K	*kappa*
λ	Λ	*lambda*
μ	M	*mu*
ν	N	*nu*
ξ	Ξ	*ksi* or *xi*
o	O	*omicron*
π or ϖ	Π	*pi*
ρ or ρ	P	*rho*
σ or ς	Σ	*sigma*
τ	T	*tau*
υ	Υ	*upsilon*
φ or ϕ	Φ	*phi*
χ	X	*khi* or *chi*
ψ	Ψ	*psi*
ω	Ω	*omega*

1.2 Various symbols

Symbol	Signification	Examples
$\exists$	*there exists, there is, there are*	Given a in $\mathbb{Z}$, then $\exists b$ in $\mathbb{Z}$ such that $b = -a$
$\forall$	*for all, for each, for every*	$\forall a, b$ elements of $\mathbb{Z}$, then $a + b$ is an element of $\mathbb{Z}$
$\mid$	*such that*	Given a in $\mathbb{Z}$, then $\exists b$ in $\mathbb{Z} \mid b = -a$
$\varnothing$	*empty set*	intersection between $]0; 1[$ and $]2; 3[$ is empty
Δ	*delta, difference*	$\lvert T_1 - T_2 \rvert = \Delta T$, difference between two temperatures
$\{...\}$	*set*	$\mathbb{P} = \{2, 3, 5, 7, 11, 13, 17, 19, ...\}$ (prime numbers)
$\in$	*is element of*	Given $a \in \mathbb{Z}$, then $\exists b \in \mathbb{Z} \mid b = -a$
$\notin$	*is not element of*	$0 \notin]1; 2[= \{x \in \mathbb{R} \mid 1 < x < 2\}$
$\subset$	*is a subset of*	$\mathbb{N} \subset \mathbb{Z} \subset \mathbb{Q} \subset \mathbb{R}$
$\not\subset$	*is not a subset of*	$\mathbb{Q} \not\subset \mathbb{Z}$, $\{1, \pi, \pi^2, \pi^3, \pi^4\} \not\subset \mathbb{Q}$
$\cup$	*union*	$A \cup B = \{x \mid x \in A \text{ or } x \in B\}$, $]0; 1[\cup [1; 2] =]0; 2]$
$\cap$	*intersection*	$A \cap B = \{x \mid x \in A \text{ and } x \in B\}$, $]0; 1[\cap [0{,}5; 2] = [0{,}5; 1[$
∞	*infinity*	$[0; +\infty[= \{x \in \mathbb{R} \mid x \geqslant 0\}$, $\lim\limits_{x \to +\infty} \frac{1}{x} = 0$
$\Rightarrow$	*implication*	Sufficient condition : $\forall a \in \mathbb{Z} \Rightarrow \exists b \in \mathbb{Z} \mid b = -a$
$\Leftrightarrow$	*if and only if*	Necessary and sufficient condition : n^2 odd $\Leftrightarrow n$ odd
$\setminus$ or $-$	*difference (set theory)*	$]0; 1[\cup]1; 2] =]0; 2] \setminus \{1\} =]0; 2] - \{1\}$

Chapter 2

Sets and Logic

2.1 Sets

2.1.1 Definition

Given a set A, the notation $x \in A$ indicates that x **is an element of** A. The **empty set** $\varnothing$ is the set that contains no element.

2.1.2 Union and Intersection

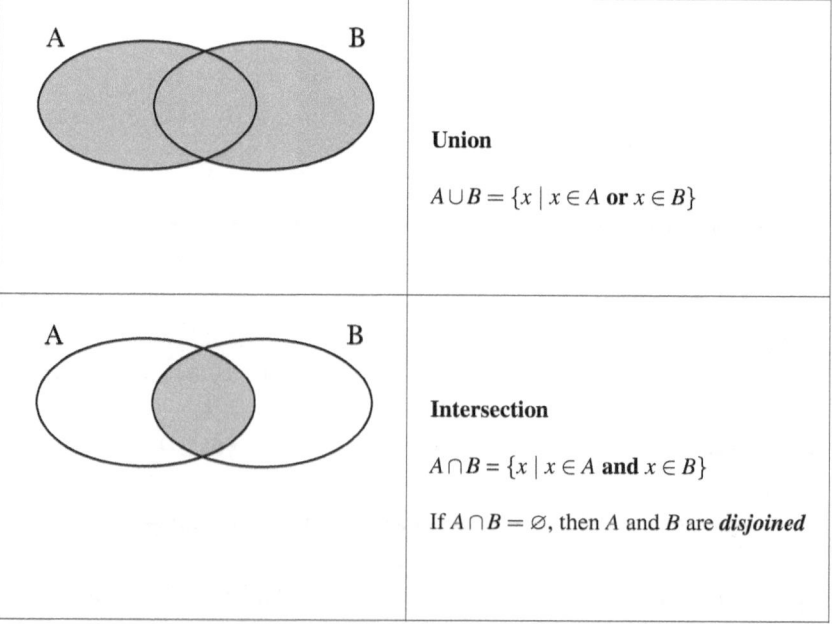

A B	**Union** $A \cup B = \{x \mid x \in A \text{ **or** } x \in B\}$
A B	**Intersection** $A \cap B = \{x \mid x \in A \text{ **and** } x \in B\}$ If $A \cap B = \varnothing$, then A and B are ***disjoined***

2.1.3   Difference and Complement

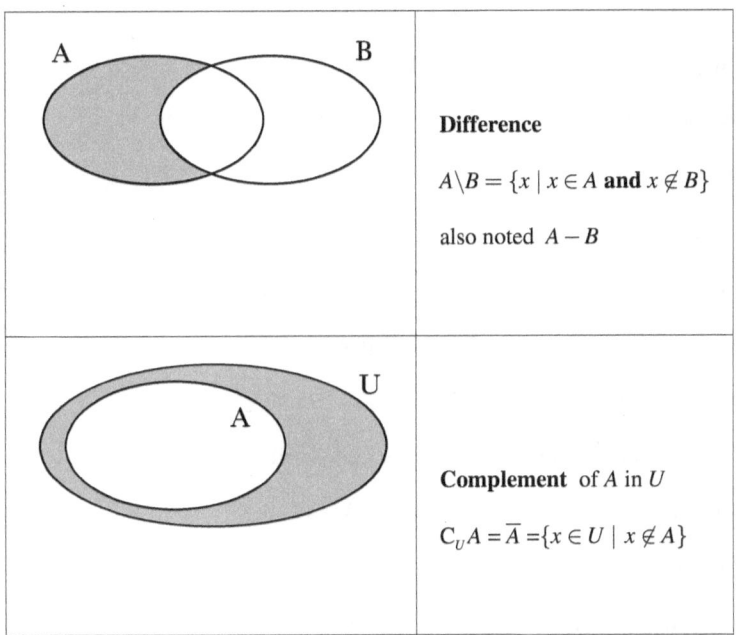

	Difference
	$A \backslash B = \{x \mid x \in A \text{ and } x \notin B\}$
	also noted $A - B$
	Complement of A in U
	$C_U A = \overline{A} = \{x \in U \mid x \notin A\}$

2.1.4   Inclusion and equality

Given three sets A, B and C. A is a subset of B if, for all $x \in A$, then $x \in B$. Notation : $A \subset B$

$(A \subset B \text{ and } B \subset A) \Leftrightarrow A = B$	$(A \subset B \text{ and } B \subset C) \Rightarrow A \subset C$

2.1.5   Properties

Commutativity	$A \cup B = B \cup A$	$A \cap B = B \cap A$
Associativity	$(A \cup B) \cup C = A \cup (B \cup C)$	$(A \cap B) \cap C = A \cap (B \cap C)$
Distributivity	$A \cup (B \cap C) = (A \cup B) \cap (A \cup C)$	$A \cap (B \cup C) = (A \cap B) \cup (A \cap C)$
Morgan's Law	$\overline{A \cup B} = \overline{A} \cap \overline{B}$	$\overline{A \cap B} = \overline{A} \cup \overline{B}$
Difference	$A \backslash A = \varnothing$	$A \backslash B = A \cap \overline{B}$

2.2 Sets of numbers

Natural numbers	$\mathbb{N} = \{0, 1, 2, 3, 4, 5, 6, ...\}$
Integers	$\mathbb{Z} = \{\ldots, -4, -3, -2, -1, 0, 1, 2, 3, 4, \ldots\}$
Rational numbers	$\mathbb{Q} = \{\frac{a}{b} \mid a \in \mathbb{Z} \text{ and } b \in \mathbb{N}\backslash\{0\}\}$
Real numbers	$\mathbb{R}$ (see 3.4.1)
Irrational numbers	$\mathbb{R}\backslash\mathbb{Q}$
Complex numbers	$\mathbb{C} = \{z = a + bi \mid a, b \in \mathbb{R} \quad \text{and} \quad i^2 = -1\}$ (see 3.4.2)
Real numbers not zero	$\mathbb{R}^* = \mathbb{R}\backslash\{0\}$
Negative real numbers	$\mathbb{R}_- = \{x \mid x \in \mathbb{R} \text{ and } x \leq 0\}$
Positive real numbers	$\mathbb{R}_+ = \{x \mid x \in \mathbb{R} \text{ and } x \geq 0\}$
Natural numbers not zero	$\mathbb{N}^* = \mathbb{N}\backslash\{0\}$
Definitions as a result	$\mathbb{Z}^*, \mathbb{Q}^*, \mathbb{Q}_-, \mathbb{Q}_+$, Etc. (impossible for $\mathbb{C}$)
Inclusion relation	$\mathbb{N} \subset \mathbb{Z} \subset \mathbb{Q} \subset \mathbb{R} \subset \mathbb{C}$

2.2.1 Intervals on $\mathbb{R}$

Closed interval	$[a; b] = \{x \in \mathbb{R} \mid a \leq x \leq b\}$
Open interval	$]a; b[= \{x \in \mathbb{R} \mid a < x < b\}$
Infinite intervals	$[a; +\infty[= \{x \in \mathbb{R} \mid a \leq x\}$
	$]-\infty; b] = \{x \in \mathbb{R} \mid x \leq b\}$
	$\mathbb{R} =]-\infty; +\infty[$

2.3   Cartesian Product of Two Sets

The Cartesian product of two sets A and B, denoted $A \times B$, is the set of all **ordered pairs** (a,b) where $a \in A$ and $b \in B$.

$$A \times B = \{(a,b)|a \in A \text{ and } b \in B\}$$

Examples :

- $A = \{1,2\}$ and $B = \{3,4\}$ $\Rightarrow$ $A \times B = \{(1,3),(1,4),(2,3),(2,4)\}$.

- $A = \{A,K,Q,J,10,9,8,7,6,5,4,3,2\}$ and $B = \{\heartsuit,\diamondsuit,\clubsuit,\spadesuit\}$. The Cartesian product of A and B is the set consisting of 52 ordered pairs, whic correspond to all 52 possible playing cards of a standart 52-card deck.

Heart Suit	(A, $\heartsuit$)	(K, $\heartsuit$)	(Q, $\heartsuit$)	(J, $\heartsuit$)	(10, $\heartsuit$)	...	(4, $\heartsuit$)	(3, $\heartsuit$)	(2, $\heartsuit$)
Diamond Suit	(A, $\diamondsuit$)	(K, $\diamondsuit$)	(Q, $\diamondsuit$)	(J, $\diamondsuit$)	(10, $\diamondsuit$)	...	(4, $\diamondsuit$)	(3, $\diamondsuit$)	(2, $\diamondsuit$)
Club Suit	(A, $\clubsuit$)	(K, $\clubsuit$)	(Q, $\clubsuit$)	(J, $\clubsuit$)	(10, $\clubsuit$)	...	(4, $\clubsuit$)	(3, $\clubsuit$)	(2, $\clubsuit$)
Spade Suit	(A, $\spadesuit$)	(K, $\spadesuit$)	(Q, $\spadesuit$)	(J, $\spadesuit$)	(10, $\spadesuit$)	...	(4, $\spadesuit$)	(3, $\spadesuit$)	(2, $\spadesuit$)

- The Cartesian product is **not commutative** : $A \times B \neq B \times A$

- The Cartesian product is **not associative** : for example, if $A = \{1\}$, then $(A \times A) \times A = \{((1,1),1)\}$ is not equal as $\{(1,(1,1)) = A \times (A \times A)\}$.

2.4   Cardinality of Sets

In mathematics, the cardinality of a **finite** set is the **number of elements of the set**. We will note $card(A)$ this number.

Examples :

- The set $A = \{1,2,3,4\}$ contains 4 elements. Therefore, $card(A) = 4$.

- The set $\mathscr{P}(A)$ of all parts of A or subsets of A, has a cardinality of $2^4 = 16 = card(\mathscr{P}(A))$. Indeed, when we build these subsets, for each element of A, we have two possible choices : to take it or not.
  $\mathscr{P}(A) = \{\{1,2,3,4\},\{1,2,3\},\{1,2,4\},\{1,3,4\},\{2,3,4\},...,\varnothing\}$.

- The totality of every water molecules on Earth is a finite set, but it's hard to put a number on its exact cardinality !

2.4.1   Countable and Uncountable Sets

Two sets A and B have the same cardinality if they are **equinumerous**, i.e. if there exists a one-to-one correspondence (a bijection) between them, i.e. if there exists a function f from A to B such that for every element y of B there is exactly one element x of A with $f(x) = y \Rightarrow card(A) = card(B)$.

These notions of cardinality and one-to-one correspondence can be extended to infinite sets like the natural numbers $\mathbb{N}$.

For example, the set $B = \{0, 2, 4, 6, 8, ...\}$ of non-negative even numbers has the same cardinality as the set $\mathbb{N}$. Here is the one-to-one correspondence. For every $y \in B$, there exists exactly one element $x \in \mathbb{N}$. That's $x = y/2$!

If there is no bijection between two sets A and B, we will say that they do not have the same cardinality and then there are only two possibilities : $card(A) < card(B)$ or $card(A) > card(B)$.

Georg Cantor published in 1891 the diagonal method, also called the **diagonalisation argument**, as a mathematical proof that $card(\mathbb{N}) < card(\mathscr{P}(\mathbb{N}))$!

On the other hand, we can easily prove that $card(\mathbb{N}) = card(\mathbb{Z}) = card(\mathbb{Q})$. The one-to-one correspondence between $\mathbb{Z}$ and $\mathbb{N}$ is as follows :

$\mathbb{N}$	0	1	2	3	4	5	6	...
$\mathbb{Z}$	0	1	-1	2	-2	3	-3	...

We said that $\mathbb{Z}$ is **countable**.

On the other hand, we can give an overview of how to show that $card(\mathbb{N}) = card(\mathbb{Q})$.

First, we must show that $card(\mathbb{N}) = card(\mathbb{N} \times \mathbb{N})$, with the correspondence :

$\mathbb{N}$	0	1	2	3	4	5	6	7	8	9	...
$\mathbb{N} \times \mathbb{N}$	(0,0)	(1,0)	(0,1)	(0,2)	(1,1)	(2,0)	(3,0)	(2,1)	(1,2)	(0,3)	...

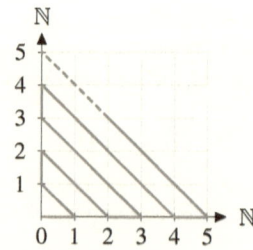

With this method (*Cantor* coupling function), we go through all the points (n,m) of the grid, where $m,n \in \mathbb{N}$, without forgetting any. So, we get to count $\mathbb{N} \times \mathbb{N}$ with $\mathbb{N}$!

Then, we define a one-to-one correspondence between $\mathbb{N} \times \mathbb{N}^*$ and $\mathbb{Q}_+$: $(a,b) \to \dfrac{a}{b}$.

A correspondence between $\mathbb{Z} \times \mathbb{N}$ and $\{1,-1\} \times (\mathbb{N} \times \mathbb{N})$ is also defined as follows : $(a,b) \to (1,(a,b))$ if $a \in \mathbb{N}$ and $(a,b) \to (-1,(a,b))$ if $-a \in \mathbb{N}^*$.

So $\mathbb{Z} \times \mathbb{N}$ is countable.

What is more, $\mathbb{Z} \times \mathbb{N}^*$ is countable and so $\mathbb{Q}$ since $card(\mathbb{N}) = card(\mathbb{N} \times \mathbb{N}^*) = card(\mathbb{Q}_+)$
$= card(\mathbb{N} \times \mathbb{N})$ $= card(\mathbb{Z} \times \mathbb{N})$ $= card(\mathbb{Z} \times \mathbb{N}^*)$ $= card(\mathbb{Q})$.

It can also be proved that the set of **algebraic numbers** ($\mathbb{A}$), which are solutions of polynomial equations with integer coefficients, is also countable, but the demonstration uses the **fundamental theorem of algebra** (of d'Alembert) and is beyond the scope of this book.

On the other hand, the set of **transcendental** numbers (non-algebraic numbers) is **uncountable** and we have : $card(\mathbb{N}) = card(\mathbb{Z}) = card(\mathbb{Q}) = card(\mathbb{A}) < card(\mathbb{R} - \mathbb{A}) = card(\mathbb{R})$.

Definitions
- Given A a set with $card(A) < card(\mathbb{N})$, we said that A is a **finite** set.
- Given B a set with $card(B) = card(\mathbb{N})$, we said that B is a **countable** set.
- Given C a set with $card(C) > card(\mathbb{N})$, we said that C is an **uncountable** set.

2.4.2   Cardinality of the Continuum

Georg Cantor showed that $card(\mathscr{P}(\mathbb{N})) > card(\mathbb{N})$ with the **diagonalisation argument,**
and then $card(\mathscr{P}(\mathbb{N})) = 2^{card(\mathbb{N})}$.

The **continuum hypothesis** states that there is no cardinal number between the cardinality of the reals and the cardinality of the natural numbers and then $card(\mathscr{P}(\mathbb{N})) = card(\mathbb{R})$.

We denoted $card(\mathbb{N}) = \aleph_0$ (aleph zero) and $card(\mathbb{R}) = \aleph_1$. Then we would have :

$$\boxed{\aleph_1 = 2^{\aleph_0}}$$

However, this hypothesis can neither be proved nor disproved.

Nevertheless, we have $card(\mathscr{P}(\mathbb{N})) < card(\mathscr{P}(\mathscr{P}(\mathbb{N}))) < card(\mathscr{P}(\mathscr{P}(\mathscr{P}(\mathbb{N}))))$, etc.

By generalizing the continuum hypothesis, $\aleph_0 < \aleph_1 < \aleph_2 < \aleph_3 < \dots$.

Yes... **there would be an infinity of infinities** ! (I know... it's very confusing!)

2.5 Logic

2.5.1 The liar paradox - Statements

A man says : « I'm lying ». If he is lying, then what he says is true and so he is not lying. If he is not lying, then what he says is true, and so he is lying. In any case, he is lying and he is not lying.

In logic and thereafter, we only use **statement** or **proposition** that can be **true or false** (*excluded third principle*) and not true and false in the same time (*non-contradiction principle*).

2.5.2 Truth tables and propositional connectives

If A is a statement, then $\neg A$ denotes the **negation** of A. Here is the truth table :

A	$\neg A$
T	F
F	T

Examples

$A : 1 + 1 = 2$ is true, $\neg A : 1 + 1 \neq 2$ is false

$A : 1 + 1 = 3$ is false, $\neg A : 1 + 1 \neq 3$ is true

When A is true, $\neg A$ is **false** (F). When A is false $\neg A$ is **true** (T).

If A and B are statements, then $A \wedge B$ denotes the **conjuncts** of A and B. Here is the truth table :

A	B	$A \wedge B$
T	T	T
T	F	F
F	T	F
F	F	F

Examples

$A : 1 + 1 = 2$ and $B : 1 + 2 = 3$ is true

$A : 1 + 1 = 2$ and $B : 1 + 2 = 2$ is false

$A : 1 + 1 = 3$ and $B : 1 + 2 = 3$ is false

$A : 1 + 1 = 3$ and $B : 1 + 2 = 2$ is false

$A \wedge B$ is true only if A and B are true at the same time (1+1=2 and 1+2=3).

If A and B are statements, then $A \vee B$ denotes the **disjuncts** of A and B. Here is the truth table :

A	B	$A \vee B$
T	T	T
T	F	T
F	T	T
F	F	F

Examples

$A : 1 + 1 = 2$ or $B : 1 + 2 = 3$ is true

$A : 1 + 1 = 2$ or $B : 1 + 2 = 2$ is true

$A : 1 + 1 = 3$ or $B : 1 + 2 = 3$ is true

$A : 1 + 1 = 3$ or $B : 1 + 2 = 2$ is false

$A \vee B$ is false only if A and B are false at the same time (1+1=3 and 1+2=2).

If A and B are statements, then $A \Rightarrow B$ denotes the **conditional** if A, then B. Here is the truth table :

A	B	$A \Rightarrow B$
T	T	T
T	F	F
F	T	T
F	F	T

Examples

if $1 + 1 = 2$, then a duck is a bird

if $1 + 1 = 2$, then a duck is a mammal

if $1 + 1 = 3$, then a duck is a bird

if $1 + 1 = 3$, then a duck is a mammal

$A \Rightarrow B$ is false when and **only when** A is true and B is false. Other cases assumed to be true.

If A and B are statements, then $A \Leftrightarrow B$ denotes the **biconditional** if A and only if B. Here is the truth table :

A	B	$A \Leftrightarrow B$		Examples
T	T	T		$1 + 1 = 2$ if and only if a duck is a bird
T	F	F		$1 + 1 = 2$ if and only if a duck is a mammal
F	T	F		$1 + 1 = 3$ if and only if a duck is a bird
F	F	T		$1 + 1 = 3$ if and only if a duck is a mammal

$A \Leftrightarrow B$ is true when and **only when** A and B have the same truth value.

The symbols $\neg$, $\wedge$, $\vee$, $\Rightarrow$ and $\Leftrightarrow$ are called **propositional connectives**.

2.5.3   Statement Forms

1. If A, B, C, A_1, B_2, Etc. are statements, every expression built up from these statement letters are **statement forms**. Examples : B, $(\neg C)$, $(A_1 \vee (\neg A_2))$, $((\neg A) \Rightarrow (\neg B))$, Etc.

2. If $\mathscr{B}$ and $\mathscr{C}$ are statement forms, then so are $(\neg \mathscr{B})$, $(\mathscr{B} \wedge \mathscr{C})$, $(\mathscr{B} \vee \mathscr{C})$, $(\mathscr{B} \Rightarrow \mathscr{C})$ and $(\mathscr{B} \Leftrightarrow \mathscr{C})$.

3. Only those expressions are statement forms that are determined to be so by mean of conditions 1 and 2 above.

For example, the statement form $((A \vee (\neg B)) \Rightarrow C)$ has the following truth table :

A	B	C	$(\neg B)$	$(A \vee (\neg B))$	$((A \vee (\neg B)) \Rightarrow C)$
T	T	T	F	T	T
T	T	F	F	T	F
T	F	T	T	T	T
F	T	T	F	F	T
T	F	F	T	T	F
F	T	F	F	F	T
F	F	T	T	T	T
F	F	F	T	T	F

2.5.4 Tautology

In logic, a tautology is a formula or statement form that is true in every possible interpretation. There are infinitely many tautologies.

Some examples :

1. $(A \lor (\neg A))$, the law of **the excluded middle**.

2. $(A \Rightarrow B) \Leftrightarrow ((\neg B) \Rightarrow (\neg A))$, the law of **contraposition**.

3. $(\neg(A \land B)) \Leftrightarrow ((\neg A) \lor (\neg B))$, the law known as **De Morgan's law**.

4. $((A \Rightarrow B) \land (B \Rightarrow C)) \Rightarrow (A \Rightarrow C)$, which is the principle known as **syllogism**.

A	B	C	$A \Rightarrow B$	$B \Rightarrow C$	$\mathscr{B} = ((A \Rightarrow B) \land (B \Rightarrow C))$	$\mathscr{C} = (A \Rightarrow C)$	$\mathscr{B} \Rightarrow \mathscr{C}$
T	T	T	T	T	T	T	T
T	T	F	T	F	F	F	T
T	F	T	F	T	F	T	T
F	T	T	T	T	T	T	T
T	F	F	F	T	F	F	T
F	T	F	T	F	F	T	T
F	F	T	T	T	T	T	T
F	F	F	T	T	T	T	T

5. $((A \Rightarrow B) \land (B \Rightarrow A)) \Leftrightarrow (A \Leftrightarrow B)$, which is the principle of **equivalence**.

A	B	$A \Rightarrow B$	$B \Rightarrow A$	$\mathscr{B} = ((A \Rightarrow B) \land (B \Rightarrow A))$	$\mathscr{C} = (A \Leftrightarrow B)$	$\mathscr{B} \Leftrightarrow \mathscr{C}$
T	T	T	T	T	T	T
T	F	F	T	F	F	T
F	T	T	F	F	F	T
F	F	T	T	T	T	T

2.5.5   Quantifiers, notations

The most common quantifiers are the **universal quantifier** $\forall$, which stands for « *for all* » or « *all* »,
and the **existential quantifier** $\exists$, which stands for « *there exists* » or « *exists* ».

Given the statement, « Each of Ryan's friends either likes to skate or likes to ride a scooter ». So, let
X be the set of all Ryan's friends, $P(x)$ the predicate « x likes to skate », and $Q(x)$ the predicate « x
likes to ride a scooter ». Then the statement can be written in formal notation as : $\forall x \in X,\ P(x) \vee Q(x)$
or even simpler, if X is well defined : $\forall x\ (P \vee Q)$.

2.5.6   Quantifiers, properties

Given X a domain of x and P, Q, **predicates** dependent on x, then we have the propreties :

Properties	Some examples or comments
$\neg(\exists x\ P) \Leftrightarrow (\forall x\ \neg P)$	no friend likes to skate $\Leftrightarrow$ all friends do not like to skate
$\neg(\forall x\ P) \Leftrightarrow (\exists x\ \neg P)$	no all friends like to skate $\Leftrightarrow$ $\exists x$ who do not likes to skate
$\neg(\forall x\ (P \Rightarrow Q)) \Leftrightarrow \exists x\ (P \wedge \neg Q)$	default of the conditional
$\neg(\exists x\ (P \wedge Q)) \Leftrightarrow \forall x\ (\neg P \vee \neg Q)$	
$\forall x\ (P \wedge Q) \Leftrightarrow ((\forall x\ P) \wedge (\forall x\ Q))$	
$(\forall x\ P) \vee (\forall x\ Q) \Rightarrow \forall x\ (P \vee Q)$	reciprocal false in general
$\exists x\ (P \vee Q) \Leftrightarrow ((\exists x\ P) \vee (\exists x\ Q))$	
$\exists x\ (P \wedge Q) \Rightarrow ((\exists x\ P) \wedge (\exists x\ Q))$	reciprocal false in general

2.5.7   Direct proof

In direct proof, the conclusion is established by logically combining the axioms, definitions, and earlier
theorems.

Consider a positive integer x, we want to show that if x is odd, then x^2 is odd, or formally :
« x is a positive odd integer » $\Rightarrow$ « x^2 is odd ». Here is the proof :

Since x is odd, it can be written $x = 2a + 1$ where a is a positive integer. Then $x^2 = (2a+1)^2 = 4a^2 + 4a + 1$. Therefore $4a^2 + 4a = 4(a^2 + a)$ is even, then $x = 4a^2 + 4a + 1$ is odd.

2.5.8   Proof by contraposition

This proof is based on the law of contraposition : $(A \Rightarrow B) \Leftrightarrow ((\neg B) \Rightarrow (\neg A))$

Consider a positive integer x, we want to show that if x^2 is even, then x is even, or formally : « x^2 is even » $\Rightarrow$ « x is even ». Here is the proof :

We need to prove the contraposal : « x is odd » $\Rightarrow$ « x^2 is odd ». But this is done in the direct proof above !

2.5.9   Proof by contradiction

Consider the Euclid's theorem : « There are infinitely many prime numbers ». Euclid offered the following proof in the $\text{III}^{\,e}$ century BC (ancient Greece).

Assume that there are only finited many primes $p_1, p_2, ..., p_r$. The number $n = p_1 \cdot p_2 \cdot ... \cdot p_r + 1$ is not divisible by any of the p_i, otherwise p_i would divide 1 the rest of the division of n by p_i for all i. This result is in contradiction with the theorem which says that any integer n can be written as a prime or as a finite product of prime numbers.

2.5.10   Proof by equivalence

We want to proof that $\overline{A \cup B} = \overline{A} \cap \overline{B}$ for any set A and B. According with the principle of equivalence, we need to proof $\overline{A \cup B} \subset \overline{A} \cap \overline{B}$ **and** $\overline{A} \cap \overline{B} \subset \overline{A \cup B}$.

For the first inclusion, given $x \in \overline{A \cup B}$, then $x \notin A \cup B$. Hence, $x \notin A$ and $x \notin B$ $\Rightarrow$ $x \in \overline{A}$ and $x \in \overline{B}$ $\Rightarrow$ $x \in \overline{A} \cap \overline{B}$, which proves $\overline{A \cup B} \subset \overline{A} \cap \overline{B}$.
For the second inclusion, given $x \in \overline{A} \cap \overline{B}$, then $x \in \overline{A}$ and $x \in \overline{B}$, hence $x \notin A$ and $x \notin B$ $\Rightarrow$ $x \notin A \cup B$ $\Rightarrow$ $x \in \overline{A \cup B}$, which proves $\overline{A} \cap \overline{B} \subset \overline{A \cup B}$.

2.5.11   Proof by mathematical induction

In a proof by mathematical induction, a single **base case** is proved, and an **induction rule** is proved that establishes that any arbitrary case implies the next case. Since in principle the induction rule can be applied repeatedly starting from the proved base case, we see that all cases are provable. This avoids having to prove each case individually.

We want to prove that the sum $1 + 2 + 3 + .. + n$ is equal to $\dfrac{n(n+1)}{2} = P(n)$ for all positive integers n.

It is true for $n = 1 = \dfrac{1(1+1)}{2}$. Assume that the result is true for $P(n-1)$ (induction rule). Then,

$$1 + 2 + 3 + ... + n - 1 = \frac{(n-1)((n-1)+1)}{2} = \frac{(n-1)n}{2} = P(n-1).$$

Hence, $1 + 2 + 3 + ... + n - 1 + n = \dfrac{(n-1)n}{2} + \dfrac{2n}{2} = \dfrac{(n-1)n + 2n}{2} = \dfrac{n((n-1)+2)}{2} = \dfrac{n(n+1)}{2}.$

2.5.12   Some conjectures

In mathematics, a conjecture is a conclusion or proposition for which no proof or disproof has yet been found.

Examples

1. **Goldbach's conjecture** : is one of the oldest and best-known unsolved problems in number theory and all of mathematics. It states: « **Every even integer greater than 2 can be expressed as the sum of two primes** » .
   Examples : $4 = 2 + 2$, $6 = 3 + 3$, $8 = 3 + 5$, $100 = 3 + 97 = 11 + 89 = 17 + 83 = 29 + 71 = 41 + 59 = 47 + 53$, Etc.

2. **The twin prime conjecture** : A twin prime is a prime number that is either 2 less or 2 more than another prime number. For example : 3 and 5, 5 and 7, 11 and 13, 17 and 19, 29 and 31, Etc.
   In other words, a twin prime is a prime that has a prime gap of two. Sometimes the term twin prime is used for a **pair of twin primes**; an alternative name for this is prime twin or prime pair. Twin primes become increasingly rare as one examines larger ranges, in keeping with the general tendency of gaps between adjacent primes to become larger as the numbers themselves get larger. However, **it is unknown whether there are infinitely many twin primes or there is a largest pair**. The work of *Yitang Zhang* in 2013, as well as work by *James Maynard*, *Terence Tao* and others, has made substantial progress towards proving that there are infinitely many twin primes, but at present this remains unsolved.

3. **My pairs of primes conjecture** : given m an even positive integer, there are infinitely many pairs of primes such that the first prime is m less than the second prime number. For example, with $m = 4$: 3 and 7, 7 and 11, 13 and 17, 19 and 23, 37 and 41, Etc.

4. **Fermat's Last Theorem** : there is no three positive integers a, b and c satisfying the equation $a^n + b^n = c^n$ for any interger value of n greater than 2. This conjecture was first conjectured by **Pierre de Fermat in 1637** in the margin of a copy of *Arithmetica*; he added that he had a proof that was too large to fit in the margin !

   The case $n = 2$ is the **Theorem of Pythagore** and it is known since antiquity to have an infinite number of solutions : $(3; 4; 5), (6; 8; 10), \dots$, $(5; 12; 13), (10; 24; 26), \dots$, $(8; 15; 17), \dots$, $(7; 24; 25), \dots$, $(20; 21; 29), \dots$, Etc. (*Pythagorean triple*).

   The first successful proof was released in 1995 by **Andrew Wiles**, after 358 years of effort by many famous mathematiciens.

   The unsolved problem stimulated the development of **algebraic number theory** in the 19th and 20th centuries and this theorem had the largest number of unsuccesful proofs until 1995.

Chapter 3

Group Theory

3.1 Closed binary operation

A **closed binary operation** on a set S is a **map** $\star$ which sends elements of the Cartesian product $S \times S$ to S.

$$S \times S \rightarrow S$$
$$(a, b) \mapsto a \star b$$

A set together with one or several closed binary operations is an **algebraic structure**.

3.1.1 Some definitions

Given $\star$ and $\diamond$ two closed binary operations on a set E, and a, b, $c \in E$:

$\star$ is **commutative** if	$a \star b = b \star a$	$\forall\, a,\, b$
$\star$ is **associative** if	$(a \star b) \star c = a \star (b \star c)$	$\forall\, a,\, b,\, c$
$e \in E$ is an **identity element** for $\star$ if	$a \star e = e \star a = a$	$\forall\, a$
$a' \in E$ is an **inverse element** of a for $\star$ if	$a \star a' = a' \star a = e$	$\forall\, a$
$\star$ is **left-distributive** over $\diamond$ if	$a \star (b \diamond c) = (a \star b) \diamond (a \star c)$	$\forall\, a,\, b, c$
$\star$ is **right-distributive** over $\diamond$ if	$(b \diamond c) \star a = (b \star a) \diamond (c \star a)$	$\forall\, a,\, b, c$

$\star$ is **distributive** over $\diamond$ if it is left-distributive and right-distributive.

The inverse a' of a is usually noted a^{-1}.

3.2 Group axioms

A **group** is a set G together with a closed binary operation called **group law**, written by example $(G, \bigstar)$, algebraic structure with the following properties :

1. $\bigstar$ **is associative**

2. **there exists an identity element** for $\bigstar$

3. **every element of** G **has an inverse** for $\bigstar$

It's easy to show, from these axioms, that **the identity element** e **is unique**.

3.2.1 Abelian group

A **group** is said **commutative** or **abelian** if, moreover, **the group law is commutative**. In that case, the group law is often denoted $+$.

Examples :

- $\mathbb{Z}$ together with the usual addition, written $(\mathbb{Z}, +)$, is an abelian group. The identity element is 0 and $-a$ is the inverse of a, $\forall a \in \mathbb{Z}$.

- $\mathbb{Q}^* = \mathbb{Q} \backslash \{0\}$ together with the usual multiplication, written $(\mathbb{Q}^*, \cdot)$, is an abelian group. The identity element is 1 and $\dfrac{1}{a}$ is the inverse of a, $\forall a \in \mathbb{Q}^*$.

- $\mathbb{Z}/_5\mathbb{Z}$ together with the addition of classes (cf. 4.2.5 modular arithmetic) is a **finite abelian group**. The identity element is $\overline{0}$ and every element has an inverse for the addition of classes.

3.2.2 Subgroup definition

A subset H of G is called a **subgroup of** G if H, together with the group law of G, also forms a group.

We will note : $(H, \bigstar) \leq (G, \bigstar)$

If $H \neq G$, we note $(H, \bigstar) < (G, \bigstar)$ and we said that $(H, \bigstar)$ is a **proper subgroup** of $(G, \bigstar)$.

The **trivial subgroup** of a group $(G, \bigstar)$ is the subgroup $(\{e\}, \bigstar)$, where e is the identity element of $(G, \bigstar)$.

Example :
$G = \mathbb{Z}$ and $H = \{2 \cdot a \mid a \in \mathbb{Z}\} =_2 \mathbb{Z}$, the subset of even integers, together with the usual addition in $\mathbb{Z}$. The identity is 0 and the inverse of an even integer in $(H, +)$ is the inverse of the even integer in $(\mathbb{Z}, +)$.

3.2.3 Subgroup properties

The identity e_H of a subgroup $(H, \bigstar)$ is the identity element of the group $(G, \bigstar)$: $e_H = e_G$.

The inverse of an element in a subgroup is the inverse of the element in the group.

A subset H of a group G is a subgroup of G **if and only if** H is **nonempty** and **closed** under the group law and inverses : $\forall a, b \in H$, then $a \bigstar b \in H$ and $a^{-1} \in H$.

These two conditions can be combined into one equivalent condition : $\forall a, b \in H$, then $(a \bigstar b)^{-1} \in H$.

3.2.4 Group homomorphisms

Given two groups $(G, *)$ and $(H, \cdot)$, a **group homomorphism** from $(G, *)$ to $(H, \cdot)$ is a **function** $h : G \mapsto H$ such that $\forall u, v \in G$, il holds that

$$h(u * v) = h(u) \cdot h(v)$$

From this property, one can deduce that $h(e_G) = e_H$ and $h(u^{-1}) = (h(u))^{-1}$.

Hence h is **compatible** with the algebraic group structure, in other words, $(H, \cdot)$ has a similar algebraic structure as $(G, *)$ and the homomorphism h ensures that.

We define the **kernel of** h : ker $(h) = \{u \in G \mid h(u) = e_H\}$, and the **image** of h : im $(h) = \{h(u) \mid u \in G\}$.

The kernel of h is a subgroup of $(G, *)$ and the **image of h is a subgroup of** $(H, \cdot)$:
by example, given $a, b \in \ker(h)$, $h(a * b) = h(a) \cdot h(b) = e_H \cdot e_H = e_H \Rightarrow (a * b) \in \ker(h)$, and if a' is such that $a * a' = e_G$, then $e_H = h(e_G) = h(a * a') = h(a) \cdot h(a') = h(a') \Rightarrow a' \in \ker(h)$.

If $h : G \mapsto H$ is such that for every $y \in H$ there exists an $x \in G$ such that $h(x) = y$, we said that h is **surjective**.

Examples :

- Given two groups $(G, *)$ and $(H, \cdot)$, we define $h : G \mapsto H$ such that $h(u) = e_H$ for all $u \in G$. Here we have $\ker(h) = G$ and im $(h) = \{e_H\}$.

- $G = \mathbb{Z}$ and $H = \mathbb{Z}$, we define $h : G \mapsto H$ such that $h(u) = -u$ for all $u \in G$. Here we have $\ker(h) = \{0\}$ and im $(h) = H$.

- $G = \mathbb{Z}$ and $H = \{2 \cdot a \mid a \in \mathbb{Z}\}$, we define $h : G \mapsto H$ such that $h(u) = 4 \cdot u$ for all $u \in G$. Here we have $\ker(h) = \{0\}$ and im $(h) = \{4 \cdot a \mid a \in \mathbb{Z}\} \subset H$. We note $H = {}_2\mathbb{Z}$ and im $(h) = {}_4\mathbb{Z}$.

3.2.5   Normal subgroup

A **normal subgroup** is a subgroup that is invariant under **conjugation** by members of the group of which it is a part. In other words, a subgroup N of the group $(G, *)$ is normal in G if and only if $g * n * g^{-1} \in N$ for all $g \in G$ and $n \in N$. We denoted then $N \triangleleft G$.

Example :
$G = (\mathbb{Z}, +)$ and $N = \{2 \cdot a \mid a \in \mathbb{Z}\} = {}_2\mathbb{Z}$, the subset of even integers.

An equivalent way to define a normal subgroup is to say that the **classes** to the right and left of N in G are the same, i.e. :

$$\forall g \in G, \ gN = \{g * n \mid n \in N\} = \{n * g \mid n \in N\} = Ng.$$

Examples :

- The trivial subgroup $\{e\}$ consisting of just the identity of G and G itself are always normal subgroups of G. If these are the only normal subgroups of G, then G is said to be **simple**.

- Every subgroup N of an abelian group G is normal, because $gN = Ng$. A group that is not abelian but for which every subgroup is normal is called a **Hamiltonian group**.

Properties :

If $N \triangleleft G$ and K a subgroup of G with $N < K$, then $N \triangleleft K$.

In contrast, if $N \triangleleft K$ and $K \triangleleft G$, N need not to be normal in G (not a transitive relation).

Two groups G and H are normal subgroups of their **direct product** $G \times H$ (analog of the Cartesian product of sets, but in algebraic structures).

If $N_1 \triangleleft G_1$ and $N_2 \triangleleft G_2$, then $N_1 \times N_2 \triangleleft G_1 \times G_2$.

If $f : G \mapsto H$ is a surjective group homomorphism and $N \triangleleft G$, then $f(N) \triangleleft H$.

If $f : G \mapsto H$ is a group homomorphism and $N \triangleleft H$, then $f^{-1}(N) \triangleleft G$.

If $N \triangleleft G$, N is the kernel of a group homorphism defined on G. Indeed, if $f : G \mapsto G'$ is a group homomorphism, $ker(f)$ is a normal subgroup of G as a reciprocal image of $\{e_{G'}\}$. Reciprocally, if $N \triangleleft G$, N is the kernel of the canonical surjection of G into the quotient group G/N (see below).

3.2.6   Quotient group

Given a group G (the law will be notated multiplicatively to simplify notations) and a normal subgroup N of G, and an element a in G, one can consider the **classe** : $aN = \{an \mid n \in N\} = \{na \mid n \in N\} = Na$

We define the set G/N as the set of all the classes aN with $a \in G$, that is :

$$G/N = \{aN \mid a \in G\}$$

We define then an operation $G/N \times G/N \mapsto G/N$ as follow : $(aN, bN) \mapsto abN$.

We have to show then that abN does not depend on the choice of the representatives, a and b, of each classe aN and bN.

To prove it, suppose $xN = aN$ and $yN = bN$ for some $x, y, a, b \in G$. Then

$abN = a(bN) = a(yN) = a(Ny) = (aN)y = (xN)y = x(Ny) = x(yN) = xyN$, since N is abelian.

It can be checked that this operation is always associative and N is the identity element of G/N.

The inverse element of aN can be represented by $a^{-1}N$.

Therefore, the set G/N together with the operation defined by $(aN)(bN) = abN$ forms a group, the **quotient group** of G by N.

Examples :

- If $N = G$, G/G has one element, the classe of $\{e_G\}$.

- If $N = e_G$, G/e_G is **isomorphic** (homomorphism bijective) to G.

- $N = {}_8\mathbb{Z} = \{..., -24, -16, -8, 0, 8, 16, 24, ...\}$ and $G = (\mathbb{Z}, +)$.
  Given $a = 2$, then $2N = \{..., -22, -14, -6, 2, 10, 18, 26, ...\}$, that we will denote $\overline{2}$.
  Given $a = 5$, then $5N = \{..., -19, -11, -3, 5, 13, 21, 29, ...\}$, that we will denote $\overline{5}$.
  Given $a = 8$, then $8N = \{..., -24, -16, -8, 0, 8, 16, 24, ...\} = N$, that we will denote $\overline{0}$, etc.
  $G/N = \mathbb{Z}/_8\mathbb{Z} = \{\overline{0}, \overline{1}, \overline{2}, \overline{3}, \overline{4}, \overline{5}, \overline{6}, \overline{7}\}$, classes modulo 8.

Properties :

The **order** of G/N, i.e. $card(G/N)$, is equal to $card(G)/card(N)$ if G is finite. The set G/N may be finite, althoug both G and N are infinite, as $\mathbb{Z}/_8\mathbb{Z}$.

There is a natural surjective group homomorphism $\pi : G \mapsto G/N$, sending each element g of G to its class, that is : $\pi(g) = gN$. This mapping π is called the **canonical surjection** of G into G/N. Its kernel is N.

There is a one-to-one correspondence between the subgroups of G that contain N and the subgroups of G/N. If $H < G$ with $N \triangleleft H$, the corresponding subgroup of H is $\pi(H)$.

3.2.7   Cyclic groups

Given a group $(G, \cdot)$, for any element g of G, one can form the subgroup of all integer powers $<g> = \{g^k \mid k \in \mathbb{Z}\}$, called the **cyclic subgroup** of g.

A **cyclic group** is a group which is equal to one of its cyclic subgroups, i.e. $G = <g>$ for some element g, called a **generator**.

We said that a cyclic group G is of **order** n if $G = \{e_G = g^0, g^1, g^2, g^3, ..., g^{n-1}\}$.

Such a group is isomorphic to the **standard cyclic group** C_n or $\mathbb{Z}/_n\mathbb{Z}$ with the addition operation.

Examples :

- The set of integers $\mathbb{Z}$, with the standard addition, forms an infinite cyclic group, because all integers can be written by repeatedly adding or substracting 1. In this group, 1 and -1 are the only generators. Every infinite cyclic group is isomorphic to $(\mathbb{Z}, +)$.

- The generators of $(\mathbb{Z}/_n\mathbb{Z}, +)$ are integers that are relatively prime (or coprime) to n. The number of such generators is $\phi(n)$ (see chapter 4.4). There are units of $(\mathbb{Z}/_n\mathbb{Z}, +)$.

- The 3th **complex roots of unity** form a cyclic group under multiplication in $\mathbb{C}$ (see 3.4.4).

- The set of **rotational symmetries** of a polygon forms a finite cyclic group. If there are n different ways of moving the polygon to itself by a rotation, then this symmetry group is isomorphic to $(\mathbb{Z}/_n\mathbb{Z}, +)$. For example, consider an equilateral triangle ABC. There are 3 possible rotations whose centre is the centre of gravity of the triangle : the identity, the one that sends A to B and the one that sends A to C.

3.3   Ring

A ring is an abelian group with a second binary operation that is associative, distributive over the abelian group operation, and has an identity element. By extension from the integers, the abelian group operation is called addition and the second binary operation is called multiplication.

Example :
The most familiar example of a ring is the set of all integers, $(\mathbb{Z}, +, \cdot)$.

Definition

A **ring** is a set R equipped with two binary operations, $+$ and $\cdot$, satisfying the following three sets of axioms, called the **ring axioms** :

1. R is an **abelian group** under addition, meaning that :

   (a) $(a+b)+c = a+(b+c)$ for all $a,b,c \in R$ (associativity)

   (b) $a+b = b+a$, for all $a,b \in R$ (commutativity)

   (c) $\exists\, 0 \in R$ such that $a+0 = 0+a = a$ for all a in R (0 is the additive identity)

   (d) $\forall a \in R$, $\exists -a \in R$ such that $a+(-a) = 0$ (additive inverse of a)

2. R is a **monoid** under multiplication, meaning that :

   (a) $(a \cdot b) \cdot c = a \cdot (b \cdot c)$ for all $a,b,c \in R$ ($\cdot$ is associative)

   (b) $\exists\, 1 \in R$ such that $a \cdot 1 = 1 \cdot a = a$ for all $a \in R$ (1 is the multiplicative identity)

3. Multiplication is **distributive with respect to addition**, meaning that :

   (a) $a \cdot (b+c) = (a \cdot b) + (a \cdot c)$ for all $a,b,c \in R$ (left distributivity)

   (b) $(a+b) \cdot c = (a \cdot c) + (b \cdot c)$ for all $a,b,c \in R$ (right distributivity)

Properties :

These basic properties follow immediately from the axioms :

- 0, the additive inverse of each element, and 1 are unique.

- For any element $x \in R$, we have $x0 = 0 = 0x$ (0 is an **absorbing element**) and $(-1)x = -x$.

- If $0 = 1$ in a ring R, then R has only one element, 0, and is called the **zero ring**.

By example, suppose that b and c are two additive inverses of a. Then, $b+a = 0$ and $c+a = 0$ $\Rightarrow$ $b+a = c+a$ $\Rightarrow$ $b = c = -a$.

Examples :

- We equip the quotient group $(\mathbb{Z}/5\mathbb{Z}, +)$ with the multiplication as follow : $\bar{a} \cdot \bar{b} = \overline{a \cdot b}$, then, $(\mathbb{Z}/5\mathbb{Z}, +, \cdot)$ is a ring. The identity for the multiplication is $\bar{1}$ and the multiplication is commutative, by property of multiplication in $\mathbb{Z}$. We said this is a **commutative ring**.

- $(\mathbb{Q}, +, \cdot)$ and $(\mathbb{R}, +, \cdot)$ are commutative rings, in particular, but we're going to see that these are also **fields** (3.4).

- The polynomial ring $R[x]$ of polynomials over a ring R is itself a ring.

3.4   Field

A **field** F is a **commutative ring** such that for every $a \neq 0 \in F$, there exists an element in F, denoted by a^{-1} or $\dfrac{1}{a}$ called the **multiplicative inverse** of a, such that $a \cdot a^{-1} = 1$.

Examples :

- $(\mathbb{Q}, +, \cdot)$ and $(\mathbb{R}, +, \cdot)$ are fields.

- $(\mathbb{Z}/_p\mathbb{Z}, +, \cdot)$ is a field if p is a prime number.

- $(\mathbb{C}, +, \cdot)$, field of Complex Numbers, is a field that contains $\mathbb{R}$ (see 3.4.2 below).

3.4.1   Field of Real Numbers + Properties

$(\mathbb{R}, +, \cdot)$ is a field that is **ordered**,
that is, there is a **total order** $\geq$ such that, for all real numbers x, y and z :

- if $x \geq y$ then $x + z \geq y + z$;

- if $x \geq 0$ and $x \geq 0$, then $xy \geq 0$.

Then we have the **system** $(\mathbb{R}, +, \cdot, \geq)$ of real numbers.

Properties :

- The order in $\mathbb{R}$ is **Dedeking-complete**. That is, every non-empty subset S of $\mathbb{R}$ with an **upper bound** in $\mathbb{R}$ has a **supremum** (least upper bound) in $\mathbb{R}$. This is not the case for $\mathbb{Q}$, example : $S = \{x \in \mathbb{Q} \mid x^2 < 2\}$ has a rational upper bound (e.g. 1.5), but no rational supremum, because, $\sqrt{2}$ is not rational.

- $\mathbb{R}$ contains all limits, i.e. if a sequence of real numbers has a limit, it is a real number. More specifically, if a sequence (x_n) of real numbers converges to the limit x, that is, if for any $\varepsilon > 0$ there exists an integer N, possibly depending on ε, such that the distance $|x_n - x|$ is less than ε for all n greater than N, then $x \in \mathbb{R}$. This means that the reals are **complete** in the sense of metric space, wich is a different sense that Dedekind completeness. Again, this is not the case for $\mathbb{Q}$.

- $\mathbb{R}$ is **Archimedian**, i.e. let $x \in \mathbb{R}$, then, there exists a natural number n such that $n > x$. Alternativaly, one can use the following logic characterization : $\forall \varepsilon \in \mathbb{R} \, (\varepsilon > 0 \Rightarrow \exists n \in \mathbb{N} \mid 1/n < \varepsilon)$

- A consequence of the fact that $\mathbb{R}$ is Archimedian is that $\mathbb{Q}$ is **dense** in $\mathbb{R}$. In other words, whatever two reals x, y such that $x < y$, there exists a rational b such that $x < b < y$. Indeed, by virtue of the foregoing, setting $\varepsilon = y - x$, there exists $n \in \mathbb{N}$ such that $1/n < \varepsilon$, then we have

$$\frac{1}{n} < y - x.$$ (1)

We define $p = \lfloor nx \rfloor$, the integer part of nx, then $p \leq nx < p+1$, by definition.

Then $\dfrac{p}{n} \leq x < \dfrac{p+1}{n}$ (2)

Summing (1) and (2) to the left of (2) and considering the right of (2),

we have finally : $x < \dfrac{p+1}{n} < y$. Then $b = \dfrac{p+1}{n}$ is the desired rational.

3.4.2 Complex Numbers

A **complex number** is a number of the form $a + bi$, where $a, b \in \mathbb{R}$ and i that is solution of the equation $x^2 = -1$. Because no real number satisfies this equation, i is called an **imaginary number**.

a is called the **real part**, and b is called the **imaginary part** of $a + bi$.

We denote $\mathbb{C}$ the set of all complex numbers.

If $a \in \mathbb{R}$, it can be regarded as a complex number $a + 0i$, hence $\mathbb{R} \subset \mathbb{C}$.

A complex number of the form $0 + bi = bi$ is called a **purely imaginary number**.

We can define a one-to-one correspondence between $\mathbb{C}$ and $\mathbb{R} \times \mathbb{R}$: $a + bi \mapsto (a, b)$.

Thus, $card(\mathbb{C}) = card(\mathbb{R} \times \mathbb{R}) = card(\mathbb{R})$ and we can visually represent $\mathbb{C}$ on a diagram, called an **Argand diagram**, representing the **complex plane** :

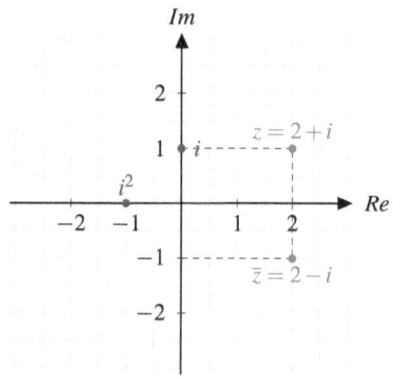

Re is the **real axis** and Im the **imaginary axis**.

$\bar{z} = a - bi$ is the **complex conjugate** of the complex number $z = a + bi$. Geometrically, $\bar{z}$ is the **reflection** of z about the real axis.

Given two complex numbers $z_1 = a_1 + b_1 i$ and $z_2 = a_2 + b_2 i$, we define, from the operation properties in $\mathbb{R}$ and $i^2 = -1$, the operations :

Addition	$z_1 + z_2 = (a_1 + a_2) + (b_1 + b_2)i$
Substraction	$z_1 - z_2 = (a_1 - a_2) + (b_1 - b_2)i$
Multiplication	$z_1 \cdot z_2 = a_1 a_2 - b_1 b_2 + (a_1 b_2 + a_2 b_1)i$

Examples :

- $(2 - 3i) + (-1 + 2i) = 2 - 1 + (-3 + 2)i = 1 - i$

- $(2 + i) - (3 - 5i) = 2 - 3 + (1 - (-5))i = -1 + 6i$

- $(3 - 5i)(-2 + i) = 3(-2) - (-5) + (3 + (-5)(-2))i = -1 + 13i$

- $(2 - i)^2 = 4 - 4i + i^2 = 3 - 4i$

- $(2 - i)(2 + i) = 4 - i^2 = 5$

Properties of the multiplication

- it is associative and commutative,
- it is distributive with respect to the addition,
- it admits the neutral element $1 + 0i$, noted 1,
- any non-zero complex number $a + bi$ admits $\dfrac{a - bi}{a^2 + b^2}$ as inverse, since $(a + bi)(a - bi) = a^2 + b^2$.

Specifically, $\dfrac{1}{a + bi} = \dfrac{a - bi}{a^2 + b^2} = \dfrac{a}{a^2 + b^2} - \dfrac{b}{a^2 + b^2} i$, hence the formula for the **division** :

$$\frac{z_1}{z_2} = \frac{a_1 + b_1 i}{a_2 + b_2 i} = \frac{a_1 a_2 + b_1 b_2}{a_2^2 + b_2^2} - \frac{a_2 b_1 - a_1 b_2}{a_2^2 + b_2^2} i$$

Examples :

- $\dfrac{1}{i} = 1 \cdot (-i) = -i$, indeed, $i \cdot (-i) = -i^2 = 1$

- $\dfrac{1 + i}{i} = (1 + i)(-i) = 1 - i$

- $\dfrac{1 + i}{1 - i} = (1 + i)\dfrac{(1 + i)}{2} = \dfrac{1 + 2i - 1}{2} = i$, indeed $(1 - i) \cdot i = 1 + i$

- $\dfrac{2 + i}{3 - 2i} = (2 + i)\dfrac{3 + 2i}{9 + 4} = \dfrac{4 + 7i}{13} = \dfrac{4}{13} + \dfrac{7}{13}i$

Uniqueness of writing complex numbers

$$a + bi = c + di \Leftrightarrow (a = c) \wedge (b = d)$$

Properties of the complex conjugate

Given z and w two complex numbers, with $w \neq 0$ for the last formula, then we have :

$z = \bar{\bar{z}}$	$\overline{z+w} = \bar{z} + \bar{w}$
$\overline{zw} = \bar{z} \cdot \bar{w}$	$\overline{(z \div w)} = \bar{z} \div \bar{w}$

3.4.3   Polar Form of the Complex Numbers

Given $z = a + bi$ a complex number, represented in the complex plane by $Z(a,b)$.

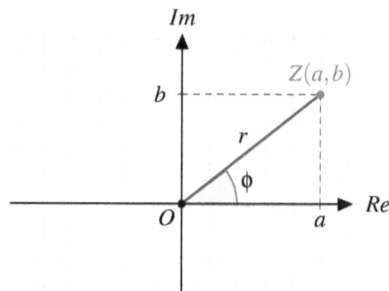

Z is entirely determined by $r = \left\| \overrightarrow{OZ} \right\| = \sqrt{a^2 + b^2}$, called **modulus** of z, or **absolute value** of z, written $|z|$, and the **argument** ϕ, the angle of the radius OZ with the positive real axis, written $\arg(z)$. If $a + bi$ is called the **rectangular form** of z, we note the **polar form** $[\, r \, , \, \phi \,]$.

We have the following relations between the rectangular and the polar form :

$\cos(\phi) = \frac{a}{r}$	$\sin(\phi) = \frac{b}{r}$
$r = \sqrt{a^2 + b^2}$	$z = r(\cos(\phi) + i \, \sin(\phi)) = [\, r \, , \, \phi \,]$

Multiplication and division in polar form

Given two complex numbers
$z_1 = r_1(\cos(\phi_1) + i\,\sin(\phi_1)) = [\,r_1\,,\,\phi_1\,]$ and $z_2 = r_2(\cos(\phi_2) + i\,\sin(\phi_2)) = [\,r_2\,,\,\phi_2\,]$,

because of the trigonometric identities
$\cos(\alpha + \beta) = \cos(\alpha)\cos(\beta) - \sin(\alpha)\sin(\beta)$ and $\sin(\alpha + \beta) = \cos(\alpha)\sin(\beta) + \sin(\alpha)\cos(\beta)$,

we have

$$z_1 z_2 = r_1 r_2(\cos(\phi_1 + \phi_2) + i\sin(\phi_1 + \phi_2)) \quad\text{and}\quad \frac{z_1}{z_2} = \frac{r_1}{r_2}(\cos(\phi_1 - \phi_2) + i\sin(\phi_1 - \phi_2))$$

or, more simply

$$[\,r_1\,,\,\phi_1\,]\cdot[\,r_2\,,\,\phi_2\,] = [\,r_1\cdot r_2\,,\,\phi_1 + \phi_2\,] \quad\text{and}\quad \frac{[\,r_1\,,\,\phi_1\,]}{[\,r_2\,,\,\phi_2\,]} = \left[\frac{r_1}{r_2}\,,\,\phi_1 - \phi_2\right]$$

Multiplications and divisions thus become much simpler than with the rectangular form since the modules are multiplied and the arguments are added !

Examples :

- $i \cdot i = \left[\,1\,,\,\dfrac{\pi}{2}\,\right] \cdot \left[\,1\,,\,\dfrac{\pi}{2}\,\right] = \left[\,1\,,\,\dfrac{\pi}{2} + \dfrac{\pi}{2}\,\right] = [\,1\,,\,\pi\,] = -1$

- $\left[\,2\,,\,\dfrac{\pi}{3}\,\right] \cdot \left[\,1.5\,,\,\dfrac{\pi}{4}\,\right] = \left[\,2\cdot 1.5\,,\,\dfrac{\pi}{3} + \dfrac{\pi}{4}\,\right] = \left[\,3\,,\,\dfrac{7\pi}{12}\,\right]$

- $\left[\,3\,,\,\dfrac{\pi}{3}\,\right] \div \left[\,1.5\,,\,\dfrac{\pi}{4}\,\right] = \left[\,2\,,\,\dfrac{\pi}{3} - \dfrac{\pi}{4}\,\right] = \left[\,2\,,\,\dfrac{\pi}{12}\,\right]$

- $\left[\,2\,,\,\dfrac{\pi}{3}\,\right]^4 = \left[\,16\,,\,\dfrac{4\pi}{3}\,\right]$

- $[\,1\,,\,\theta\,]^n = [\,1\,,\,n\theta\,]$, it's the **Moivre's formula**

- $[\,2\,,\,\pi/8\,]^{10} = [\,1024\,,\,5\pi/4\,]$

- $\left[\,\sqrt{2}\,,\,\pi/3\,\right]^6 = [\,8\,,\,2\pi\,] = [\,8\,,\,0\,]$

3.4.4 n th Roots of a Complex Number

$$\sqrt[n]{[\, r,\, \phi\,]} = \left[\, \sqrt[n]{r},\, \frac{\phi + 2k\pi}{n}\, \right] \text{, for } 0 \leq k \leq n-1 \text{ and } k \in \mathbb{N}$$

Examples :

- $\sqrt[3]{1} = \sqrt[3]{[\, 1,\, 0\,]} = \left[\, 1,\, \dfrac{0 + 2k\pi}{3}\, \right]$, for $k = 0, 1, 2$,

  then there are three roots : $z_0 = [\, 1,\, 0\,]$, $z_1 = \left[\, 1,\, \dfrac{2\pi}{3}\, \right]$, $z_2 = \left[\, 1,\, \dfrac{4\pi}{3}\, \right]$

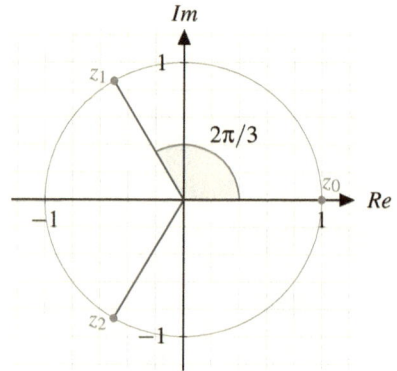

This is the cyclic group of order 3 we talk about in (3.2.7), $G = \{z_0, z_1, z_2\}$ with the multiplication in $\mathbb{C}$, $z_0 = 1$ is the identity and z_1 and z_2 are each generator.

- $\sqrt[4]{i} = \sqrt[4]{[\, 1,\, \pi/2\,]} = \left[\, 1,\, \dfrac{\pi/2 + 2k\pi}{4}\, \right]$, for $k = 0, 1, 2, 3$,

  then there are four roots : $z_0 = [\, 1,\, \pi/8\,]$, $z_1 = \left[\, 1,\, \dfrac{5\pi}{8}\, \right]$, $z_2 = \left[\, 1,\, \dfrac{9\pi}{8}\, \right]$, $z_3 = \left[\, 1,\, \dfrac{13\pi}{8}\, \right]$.

  But it's not a cyclic group because the identity does not exist.

- $\sqrt{[\, 4,\, 3\pi/2\,]} = \left[\, 2,\, \dfrac{3\pi/2 + 2k\pi}{2}\, \right]$, for $k = 0, 1 \Rightarrow z_0 = [\, 2,\, 3\pi/4\,]$, $z_1 = \left[\, 2,\, \dfrac{7\pi}{4}\, \right]$

  These two roots are symmetrical in the complex plane with respect to $(0,0)$.

- $\sqrt{1+i} = \sqrt{[\, \sqrt{2},\, \pi/4\,]} = \left[\, \sqrt[4]{2},\, \dfrac{\pi/4 + 2k\pi}{2}\, \right] \Rightarrow z_0 = [\, \sqrt[4]{2},\, \pi/8\,]$, $z_1 = \left[\, \sqrt[4]{2},\, \dfrac{9\pi}{8}\, \right]$

  These two roots are symmetrical in the complex plane with respect to $(0,0)$.

3.4.5   Algebraic Properties of The Field of Complex Numbers

With the properties of the multiplication and the addition, $\mathbb{C}$ is a field, called the **field of complex numbers**.

On the other hand, $\mathbb{C}$ **is not an ordered field**, unlike $\mathbb{R}$.

Solving complex second-degree equations

The equation $az^2 + bz + c = 0$, with $a, b, c \in \mathbb{C}$ has two complex solutions, with multiplicity or not.

$$z_1 = \frac{-b + \sqrt{b^2 - 4ac}}{2a} \quad \text{and} \quad z_2 = \frac{-b - \sqrt{b^2 - 4ac}}{2a}$$

Example :

$z^2 - (2 + 3i)z - (5 - i) = 0$, then $b^2 - 4ac = (2 + 3i)^2 + 4(5 - i) = 15 + 8i = (4 + i)^2$

$\Rightarrow z_1 = \dfrac{2 + 3i + 4 + i}{2} = 3 + 2i$ and $z_2 = \dfrac{2 + 3i - 4 - i}{2} = -1 + i$

One can check, $(z - (3 + 2i))(z - (-1 + i)) = z^2 - 2z - 3iz - 5 + i = z^2 - (2 + 3i)z - (5 - i)$.

Fundamental Theorem of Algebra

> Any polynomial with complex coefficients of degree n ($n \geq 1$) is factorizable into a product of n first-degree factors with complex coefficients.

Because of this fact, the field of complex numbers $\mathbb{C}$ is called an **algebraically closed field**.

This property does not hold for $\mathbb{Q}$ nor $\mathbb{R}$:
- the polynomial $x^2 - 2$ does not have a rational root in $\mathbb{Q}$
- the polynomial $x^2 + 1$ does not have a real root in $\mathbb{R}$

However, for R, we have the following theorem that follows from the previous theorem :

> Any polynomial with real coefficients of degree n ($n \geq 1$) is factorizable into a product of factors with real coefficients that are first or second degree irreductible in $\mathbb{R}$.

In particular, if n is odd, we can be sure that there is at least one real root.

Example :

$z^3 - 6z^2 + 13z - 10 = (z - 2)(z^2 - 4z + 5)$, with $z^2 - 4z + 5$ not factorizable in $\mathbb{R}$ but with 2 complex roots conjugate in $\mathbb{C}$:

$b^2 - 4ac = -4 = (2i)^2 \Rightarrow w = 2 + i$ and $\overline{w} = 2 - i$, thus $(z - (2 + i))(z - (2 - i)) = z^2 - 4z + 5$.

3.4.6 Complex Exponential Function

$\mathbb{C}$, with the **metric** : $d(z_1, z_2) = |z_2 - z_1|$, is a **complete metric space**, which notably includes the **triangle inequality** $|z_1 + z_2| \leq |z_1| + |z_2|$ for any two complex numbers z_1 and z_2.

Like in analysis, we define notions of convergent series from this metric. This notion is used to construct a number of elementary functions including the **exponential function** $\exp(z)$, also written e^z, and defined as the infinite convergent series

$$\exp(z) = 1 + z + \frac{z^2}{2 \cdot 1} + \frac{z^3}{3 \cdot 2 \cdot 1} + \ldots = \sum_{n=0}^{\infty} \frac{z^n}{n!}$$

If $z = 1$, the series converges to the **Euler number** : $e = 2.718, 281, 828, 459, \ldots$

If $z = 0$, we have $\exp(0) = 1$.

Using the fact that $i^0 = 1$, $i^1 = i$, $i2 = -1$, $i3 = -i$, $i^4 = 1$, etc. and the **Maclaurin** series for $\cos(x)$ and $\sin(x)$, Euler showed that

$$\exp(i\phi) = \cos(\phi) + i\sin(\phi)$$

Hence Euler's magnificent formula linking e, π and i :

$$e^{i\pi} = -1$$

Chapter 4

Number Theory

4.1   Prime numbers

4.1.1   Divisibility

Given a non-zero integer m, we say that m **divides** an integer n, or that n **is divisible by** m, if there exists an integer r such as $n = r \cdot m$. We also say that m is a **divisor** of n or although that n is a **multiple** of m. Written $m \mid n$. Otherwise, it's written $m \nmid n$.

4.1.2   Properties

1. if $m \mid n$ and $m \mid r$, then $m \mid (n+r)$ and $m \mid (n-r)$

2. if $m \mid n$ and $r \in \mathbb{Z}$, then $m \mid r \cdot n$

3. if r, m, $n \in \mathbb{Z}$, and $r \neq 0$, then $m \mid n \Leftrightarrow r \cdot m \mid r \cdot n$

4. if $m \mid n$ and $n \mid r$, then $m \mid r$ (*transitivity*)

5. every non-zero integer m divides 0, because $m \cdot 0 = 0$

4.1.3   Divisibility criteria

an integer is divisible by...	if
2	it is even
5	it ends with 0 or 5
3	the sum of its digits is divisible by 3
6	it is divisible by 2 and by 3
9	the sum of its digits is divisible by 9
11	the sum of odd rank digits subtracted from the sum of even rank digits is divisible by 11

4.1.4   Prime number definition

A **prime number** is a positive integer which possesses exactly two divisors : 1 and itself.
The smallest prime is $2 = 1 \cdot 2$, the only even prime number, then we have 3, 5, 7, 11, 13, 17, 19, ...

4.1.5   Fundamental Theorem of Arithmetic

> Every integer $n > 1$ is either prime itself or is the product of prime numbers, and that,
>
> although the order of the primes is arbitrary in the second case, the primes themselves are not

Examples :
$$36 = 2 \cdot 2 \cdot 3 \cdot 3 = 3 \cdot 2 \cdot 2 \cdot 3 = 2^2 \cdot 3^2, \ 51 = 3 \cdot 17, \ 100 = 2^2 \cdot 5^2, \ 1001 = 7 \cdot 11 \cdot 13, \ 1111 = 11 \cdot 101$$

The proof can be made by induction, but the uniqueness need the **Gauss Lemma** we shall see below.

4.1.6   Large prime numbers

The largest known prime number (as of January 2019) is $2^{82,589,933} - 1$, a number which has $24,862,048$ digits when written in base 10. It was found by *Patrick Laroche* of the *Great Internet Mersenne Prime Search* (GIMPS) in 2018.

Modern **cryptology** is based mainly on the difficulty of factoring large integers containing a product of two large primes (RSA). Systems in the most secure communication are based on *key* with 2048 bits. This means that a person who wishes to decode a message encrypted with such a key should factorize a number which contains 2048 base 2 digits (1 or 0). This number has 617 digits written in base 10.

4.1.7   Mersenne prime numbers

In mathematics, a **Mersenne prime** is a prime number that is one less than a power of two. That is, it is a prime number of the form $M_p = 2^p - 1$ for some integer p (that need to be prime, but it is not sufficient). This is the list of the top ten Mersenne primes :

p	M_p	M_p digits	*discovered*
2	3	1	Ancient Greek mathematicians
3	7	1	Ancient Greek mathematicians
5	31	2	Ancient Greek mathematicians
7	127	3	Ancient Greek mathematicians
13	8191	4	1456
17	131,071	6	1588 by *Pietro Cataldi* (IT)
19	524,287	6	1588 by *Pietro Cataldi*
31	2,147,483,647	10	1772 by *Leonhard Euler* (CH)
61	2,305,843,009,213,693,951	19	1883 by *Ivan M. Pervushin* (RUS)
89	618,970,019,642,690,137,449,562,111	27	1911 June by *Ralph Ernest Powers* (US)

4.1.8   Perfect numbers

The Mersenne numbers are related to **perfect numbers** that have the property of being equal to the sum of their **proper divisors**.

Euclid had already shown that if $M = 2^p - 1$ is prime, then $M(M+1)/2 = 2^{p-1}(2^p - 1)$ is a perfect number.

M_p	$M(M+1)$	proper divisors	sum
$M_2 = 3$	$6 = 2 \cdot 3$	1; 2; 3	6
$M_3 = 7$	$28 = 2^2 \cdot 7$	1; 2; 4; 7; 14	28
$M_5 = 31$	$496 = 2^4 \cdot 31$	1; 2; 4; 8; 16; 31; 62; 124; 248	496
$M_7 = 127$	$8128 = 2^6 \cdot 127$	1; 2; 4; 8; 16; 32; 64; 127; 254; 508; 1016; 2032; 4064	8128

4.1.9   Euclid's Theorem

As shown in the 2.4.9 subsection, the Euclid's theorem says that

> **there are infinitely many prime numbers**

4.1.10   Prime numbers theorem

In 1737, *Euler* analysis has reached the following conclusions :

1. prime numbers are distributed irregularly (there is no formula)

2. sometimes they are very close (twin primes), sometimes they are very far

3. prime numbers are fewer and fewer

Gauss noticed in 1792 (at age 15!) That the primes are distributed according to a logarithmic law, but could not demonstrate. The proof of this result was one of the biggest challenges for mathematicians of the nineteenth century. It is the following theorem.

> The number of primes up to n is approximately given by $\dfrac{n}{\ln(n)}$

In other words, the probability that a random integer n is prime is approximately equal to $\dfrac{1}{\ln(n)}$.

4.1.11 Titanic primes

Titanic prime is a term coined by *Samuel Yates* in the 1980s, denoting a prime number of **at least 1000 decimal digits**. Few such primes were known then, but the required size is trivial for modern computers.

The **first discovered titanic primes** were the Mersenne primes $M_{4253} = 2^{4253} - 1$ (with 1281 digits), and $M_{4423} = 2^{4423} - 1$ (with 1332 digits). They were both found November 3, 1961, by *Alexander Hurwitz*. And the surprising, M_{4423} was computed first with an IBM 7090.

Here are the 5 first titanic Mersenne primes :

p	M_p	M_p digits	discovered
4,253	$2^{4,253} - 1$	1,281	1961 Nov. by *Alexander Hurwitz*
4,423	$2^{4,423} - 1$	1,332	1961 Nov. by *Alexander Hurwitz*
9,689	$2^{9,689} - 1$	2,917	1963 May 11 by *Donald B. Gillies*
9,941	$2^{9,941} - 1$	2,993	1963 May 16 by *Donald B. Gillies*
11,213	$2^{11,213} - 1$	3,376	1963 June 2 by *Donald B. Gillies*

The **first 10 titanic primes** are of the form : $p = 10^{999} + n$, for $n = 7$, 663, 2121, 2593, 3561, 4717, 5863, 9459, 11239, 14397 and can be easily tested by modern computers in few seconds with *Python, Mupad, Wolfram Alpha*, etc. (command : *isprime()*). And for the beauty of mathematics :

$10^{999} + 7 =$

100000000000000000000000000000000000000000000000000000000000000000000000000000
000000000000000000000000000000000000000000000000000000000000000000000000000000
000000000000000000000000000000000000000000000000000000000000000000000000000000
000000000000000000000000000000000000000000000000000000000000000000000000000000
000000000000000000000000000000000000000000000000000000000000000000000000000000
000000000000000000000000000000000000000000000000000000000000000000000000000000
000000000000000000000000000000000000000000000000000000000000000000000000000000
000000000000000000000000000000000000000000000000000000000000000000000000000000
000000000000000000000000000000000000000000000000000000000000000000000000000000
000000000000000000000000000000000000000000000000000000000000000000000000000000
000000000000000000000000000000000000000000000000000000000000000000000000000000
000000000000000000000000000000000000000000000000000000000000000000000000000000
0000000000000000000000000000000000000000000000000000000007

4.1.12 Gigantic primes

A **gigantic prime** is a prime number with **at least 10,000 decimal digits**. The term appeared in *Journal of Recreational Mathematics* in the article *Collecting gigantic and titanic primes* (1992) by *Samuel Yates*.

The **first discovered gigantic prime** was the Mersenne prime $2^{44,497} - 1$. It has 13,395 digits and was found in 1979 by *Harry L. Nelson* and *David Slowinski*.

The **smallest gigantic prime** is $10^{9,999} + 33,603$. It was proved prime in 2003 by *Jens Franke, Thorsten Kleinjung* and *Tobias Wirth* with their own distributed *elliptic curve primality program* (ECPP).

4.1.13   Megaprimes

A **megaprime** is a prime number with at least one million decimal digits.

The first to be found was the Mersenne prime $2^{6,972,593} - 1$ with 2,098,960 digits, discovered in 1999 by *Nayan Hajratwala*, a participant in the distributed computing project GIMPS.

In fact, **almost all primes are megaprimes**, as the number of primes with less than a million digits is finite. However, the vast majority of known primes are not megaprimes.

We don't know yet twin megaprime, as **the current largest twin prime known** is $2,996,863,034,895 \cdot 2^{1290000} \pm 1$ with *only* 388,342 decimal digits ! It was discovered in September 2016. To set an order of magnitude, there are 808,675,888,577,436 twin prime pairs below 10^{18} .

4.2   Congruence and GCD

4.2.1   Euclidean division in $\mathbb{Z}$

Given a and b integers with $b \neq 0$.

There exists an unique pair of integers (q, r), such that $a = b \cdot q + r$, and $0 \leq r < b$

This is the principle of distribution of a identical objects to b persons, so that each person receives the maximum q objects and equitable manner, the remainder r, smaller than the number of persons, is not distributed.

Example
$a = 50$, $b = 8$, $50 = 8 \cdot 6 + 2$, let's note we have too $50 = 8 \cdot 5 + 10$, but the remainder is greater than the divisor.

4.2.2   Congruence modulo n

Given a positive integer $n > 1$.
Two integers a and b are called **congruent modulo** n if n divides $(a - b)$. Written $a \equiv b \ (\bmod n)$.

Examples
$2 \equiv 12 \ (\bmod 5)$, because 5 divides $2 - 12 = -10$.
$2 \equiv -1 \ (\bmod 3)$, because 3 divides $2 - (-1) = 3$.
$49 \equiv 0 \ (\bmod 7)$, because 7 divides $49 - 0 = 49$.
$1001 \equiv 1 \ (\bmod 10)$, because 10 divides $1001 - 1 = 1000$.

4.2.3 Properties of the congruence relation

The congruence relation has following properties in $\mathbb{Z}$:

1. $a \equiv a \ (\ \mathrm{mod}\ n\ )$ **(reflexivity)**

2. if $a \equiv b \ (\ \mathrm{mod}\ n\ )$, then $b \equiv a \ (\ \mathrm{mod}\ n\ )$ **(symmetry)**

3. if $a \equiv b \ (\ \mathrm{mod}\ n\ )$ and $b \equiv c \ (\ \mathrm{mod}\ n\ )$, then $a \equiv c \ (\ \mathrm{mod}\ n\ )$ **(transitivity)**

4. an integer a is congruent modulo n to exactly one number in the set $\{0,1,2,...,n-2,n-1\}$,
   and this number is the remainder of the Euclidean division of a by n.
   The set of integers $\{0,1,2,...,n-2,n-1\}$ is called the **least residue system** modulo n.

The first three properties indicate that the congruence relation is an **equivalence relation**.

4.2.4 Congruence classes

The set $\{...,a-2n,a-n,a,a+n,a+2n,...\}$, denoted $\bar{a}_n$ is **the equivalence class** of the integer a, modulo n, also called **residue class** of the integer a, modulo n.

Example :
$$\{...,-5,-2,1,4,7,10,...\} = \bar{1}_3 = \bar{7}_3 = ...$$

Each residue class modulo n may be represented by any one of its members, although we usually represent each residue class by the smallest nonnegative integer which belongs to that class. Furthermore, **every integer belongs to one and only one residue class** modulo n.

Example :
$$\{...,-6,-3,0,3,6,9,...\} = \bar{0}_3$$
$$\{...,-5,-2,1,4,7,10,...\} = \bar{1}_3$$
$$\{...,-4,-1,2,5,8,11,...\} = \bar{2}_3$$

The set of all congruence classes of the integers for a modulus n is called the **ring of integers modulo** n, and is denoted $\mathbb{Z}/n\mathbb{Z}$ (see 3.3).

Examples :
$$\mathbb{Z}/3\mathbb{Z} = \{\bar{0}_3, \bar{1}_3, \bar{2}_3\}$$
$$\mathbb{Z}/5\mathbb{Z} = \{\bar{0}_5, \bar{1}_5, \bar{2}_5, \bar{3}_5, \bar{4}_5\}$$
$$\mathbb{Z}/2\mathbb{Z} = \{\bar{0}_2, \bar{1}_2, \}$$

4.2.5 Modular arithmetic

We defined addition, substraction, and multiplication on $\mathbb{Z}/n\mathbb{Z}$ by the following rules, based on the properties of the operations in $\mathbb{Z}$:

$$\bar{a}_n + \bar{b}_n = \overline{(a+b)}_n \quad | \quad \bar{a}_n - \bar{b}_n = \overline{(a-b)}_n \quad | \quad \bar{a}_n \cdot \bar{b}_n = \overline{(a \cdot b)}_n$$

In this way, $\mathbb{Z}/n\mathbb{Z}$ becomes a **commutative ring** (the multiplication operation is commutative).

Examples :

- multiplication table in $\mathbb{Z}/5\mathbb{Z}$

$\cdot$	$\bar{0}$	$\bar{1}$	$\bar{2}$	$\bar{3}$	$\bar{4}$
$\bar{0}$	$\bar{0}$	$\bar{0}$	$\bar{0}$	$\bar{0}$	$\bar{0}$
$\bar{1}$	$\bar{0}$	$\bar{1}$	$\bar{2}$	$\bar{3}$	$\bar{4}$
$\bar{2}$	$\bar{0}$	$\bar{2}$	$\bar{4}$	$\bar{1}$	$\bar{3}$
$\bar{3}$	$\bar{0}$	$\bar{3}$	$\bar{1}$	$\bar{4}$	$\bar{2}$
$\bar{4}$	$\bar{0}$	$\bar{4}$	$\bar{3}$	$\bar{2}$	$\bar{1}$

We can see that every non-zero element has an inverse for the multiplication. In this case, that's because $n = 5$ is prime and we said that $\mathbb{Z}/5\mathbb{Z}$ is a **finite field**.

- addition table in $\mathbb{Z}/5\mathbb{Z}$

$+$	$\bar{0}$	$\bar{1}$	$\bar{2}$	$\bar{3}$	$\bar{4}$
$\bar{0}$	$\bar{0}$	$\bar{1}$	$\bar{2}$	$\bar{3}$	$\bar{4}$
$\bar{1}$	$\bar{1}$	$\bar{2}$	$\bar{3}$	$\bar{4}$	$\bar{0}$
$\bar{2}$	$\bar{2}$	$\bar{3}$	$\bar{4}$	$\bar{0}$	$\bar{1}$
$\bar{3}$	$\bar{3}$	$\bar{4}$	$\bar{0}$	$\bar{1}$	$\bar{2}$
$\bar{4}$	$\bar{4}$	$\bar{0}$	$\bar{1}$	$\bar{2}$	$\bar{3}$

4.2.6   Greatest Common Divisor - $GCD(a,b)$

Ten GCD of two integers a and b is their greatest common divisor. Written $GCD(a,b)$.
Here are some important properties of the GCD of two positive integers a and b :

1. $GCD(a,a) = a$

2. $GCD(a,0) = a$

3. $GCD(a,b) = GCD(a-b,b)$, if $a > b$

4. $GCD(a,b) = GCD(r,b)$, where r is the remainder of the division of a by b

5. $GCD(a,b) = GCD(b,a)$

We can prove the fourth property : by Euclidean division of a by b, there are integers q and r, uniquely
determined, such that $a = b \cdot q + r$, with $0 \leq r < b$.
If $d \mid a$ and $d \mid b$, then $d \mid bq$. Thus $d \mid a - bq = r$.
Conversly, if $d \mid r$ and $d \mid b$, then $d \mid a = bq + r$, hence the conclusion.

These properties make it possible to calculate the GCD of two numbers.
Examples :
- $GCD(60,36) = GCD(60-36,36) = GCD(24,36) = GCD(36,24) = GCD(12,24) = 12$
- $GCD(1320,840) = GCD(840,480) = GCD(480,360) = GCD(360,120) = GCD(120,0) = 120$

4.2.7   Euclid's algorithm

```
while (b != 0) // while b is not equal to 0
{
  r = a % b // remainder of Euclidean division of a by b
  a = b
  b = r
}
GCD=a // last non-zero remainder
```

Using the fourth property 4.2.6, we can replace a by b and b by r without changing the GCD. We
can repeat this process until r is equal to zero and goes back from one iteration to get the last non-zero
remainder that divides, consequently, the penultimate non-zero remainder.

This necessarily happens after a finite number of iterations since the remainder, at each step, is strictly
smaller than the previous.

Finally, we'll get :
$GCD(a,b) = GCD(b,r) = ... = GCD(r_{n-2},r_{n-1}) = GCD(r_{n-1},0) = r_{n-1}$ ($r_n = 0$)
This algorithm is incredibly simple and efficient, far more than what schoolchildren learn when they
search for all the common divisors of two numbers to find the largest one (last example above).

4.2.8   GCD Theorem

Given a and b positive integers. Then there exists integers x and y such that $a \cdot x + b \cdot y = GCD(a,b)$
The proof of this theorem uses an algorithm known as the **extended Euclidean algorithm**.
One start with the system of equations :

$$\begin{bmatrix} a &=& 1 \cdot a &+& 0 \cdot b \\ b &=& 0 \cdot a &+& 1 \cdot b \end{bmatrix} \tag{4.1}$$

applying the following linear combination, the first equation multiplied by 1 added to the second multiplied by $(-q)$, where q is the quotient of Euclidean division of a by b : $a = q \cdot b + r$, with $0 \le r < b$.
We obtain, since $r = a - q \cdot b$:

$$\begin{bmatrix} a &=& 1 \cdot a &+& 0 \cdot b \\ r &=& 1 \cdot a &+& (-q) \cdot b \end{bmatrix} \tag{4.2}$$

we repeat this process on the system :

$$\begin{bmatrix} b &=& 0 \cdot a &+& 1 \cdot b \\ r &=& 1 \cdot a &+& (-q) \cdot b \end{bmatrix} \tag{4.3}$$

there exists q_1 and r_1 such that $b = q_1 \cdot r + r_1$, with $r_1 < r$, hence the same linear combination as above with q_1 instead of q.

$$\begin{bmatrix} b &=& 0 \cdot a &+& 1 \cdot b \\ r_1 &=& (-q_1) \cdot a &+& (-q_1)(-q) \cdot b + b \end{bmatrix} \tag{4.4}$$

we repeat the process on the system :

$$\begin{bmatrix} r &=& 1 \cdot a &+& (-q) \cdot b \\ r_1 &=& (-q_1) \cdot a &+& (1 + (-q_1)(-q)) \cdot b \end{bmatrix} \tag{4.5}$$

At each step, (4.3) and (4.5), the left side of the second equation is strictly inferior than the left side of the first.

With this condition, we know that, after a finite numbre of iterations, the left side of the second equation will be zero.

Then we'll have the $GCD(a,b)$ to the penultimate step equal to the last non-zero remainder, and the numbers x and y that we sought at the right side.

Let's look at an *example* :

Line	r	x	y	Comment	q
1	$1320 = a$	1	0		
2	$280 = b$	0	1	$1320 = 4 \cdot 280 + 200$	$4 = q$
3	$200 = r$	1	-4	$280 = 1 \cdot 200 + 80$	$1 = q_1$
4	$80 = r_1$	-1	5	$200 = 2 \cdot 80 + 40$	$2 = q_2$
5	$\mathbf{40 = r_2}$	$\mathbf{3}$	$\mathbf{-14}$	$80 = 2 \cdot 40 + 0$	$2 = q_3$
6	$0 = r_3$			end algorithm	

Thus, we have $GCD(1320, 280) = 40 = 3 \cdot 1320 + (-14) \cdot 280$, all right !

Corollary :

Given a, b and d integers. If d divides a and b, then d divides $GCD(a,b)$.

As matter of fact, if there exists u and v such that $d \cdot u = a$ and $d \cdot v = b$, there exists integers x and y such that $ax + by = GCD(a,b)$. Whence $dux + dvy = d(ux + vy) = GCD(a,b)$.

4.2.9   Gauss Lemma

Given a, b and c integers, if a divides $b \cdot c$ and if $GCD(a,b) = 1$, then a divides c.

By the theorem GCD, there exists integers x and y such that $ax + by = 1$. Since $a \mid b \cdot c$, there exists u integer such that $a \cdot u = b \cdot c$. Then, $ax + by = 1 \Rightarrow acx + bcy = c \Rightarrow acx + auy = c \Rightarrow a(cx + uy) = c$, this proves that a divides c.

4.2.10   Unique factorization of an integer

We are able, this time, to demonstrate the **uniqueness** of the factorization of an integer n into a product of prime factors (4.1.5).

First, we must show by induction, that if a prime p divides a product of m factors, then it divides one of them. It's a consequence of Gauss Lemma.

Next, assume that $n = p_1 \cdot p_2 \cdot \ldots \cdot p_k = q_1 \cdot q_2 \cdot \ldots \cdot q_r$ where every p_i and q_j are prime for $i = 1, \ldots, k$ and $j = 1, \ldots, r$ (there can be several times the same prime factor, of course).

Then p_1 divides one of the q_j by the above. Even if we rearrange the q_j, let's assume that $p_1 \mid q_1$.

Hence, $p_1 = q_1$ since $GCD(p_1, q_1) = 1$.

We continue the same reasoning with p_2 and so on, until every p_i il exactly a q_i. Finally, we will also have $k = r$ and n decomposes into a unique product of prime factors $p_1 \cdot p_2 \cdot \ldots \cdot p_k$.

Examples : $6{,}409 = 13 \cdot 17 \cdot 29$ and $3{,}307{,}500 = 2^2 \cdot 3^3 \cdot 5^4 \cdot 7^2$, whereas $1{,}234{,}567{,}891$ is prime.

4.3 Modular algebra

We have seen in 4.2.5 that if $a \equiv c \ (\mod n)$ and $b \equiv d \ (\mod n)$, then $ab \equiv cd \ (\mod n)$.

The reciprocal is false in general. Example : $2 \cdot 5 \equiv 2 \cdot 9 (\mod 8)$, but $5 \not\equiv 9 (\mod 8)$!

We need the following theorem to clarify the situation.

Theorem

> Given a, b, k and n integers, with $n > 1$ and $k \neq 0$.
>
> Then $ka \equiv kb \ (\mod n)$ **and** $GCD(k,n) = 1 \Rightarrow a \equiv b \ (\mod n)$

With this theorem, that uses Gauss Lemma, we can solve congruence of the type : $ax \equiv b \ (\mod n)$, provided that $GCD(a,n) = 1$.

But we need an intermediate **proposition** :

> $ax \equiv 1 \ (\mod n)$ admits a solution $\Leftrightarrow GCD(a,n) = 1$

Remark : This result gives, in particular, the existence condition for an **inverse** of a modulo n. We call **unit** of $\mathbb{Z}/_n\mathbb{Z}$ such an element and we'll denote its inverse a^{-1}.

Demonstration :
($\Rightarrow$) : if $ax \equiv 1 \ (\mod n)$, there exists integer y such that $ny = ax - 1$. Then $n(-y) + ax = 1 \Rightarrow GCD(a,n) = 1$ (every divisor of a and n divides 1).
($\Leftarrow$) : by theorem GCD, there exists integers x and y such that $ny + ax = 1 = GCD(a,n)$
$\Rightarrow n \,|\, ax - 1 \Rightarrow ax \equiv 1 \ (\mod n)$.

4.3.1 Linear congruence $ax \equiv b (\mod n)$

Theorem

> If $GCD(a,n) = 1$, the congruence $ax \equiv b \ (\mod n)$ has a solution for every b

Demonstration :
By the previous proposition, $ax \equiv 1 \ (\mod n)$ has a solution u. Then there exists an integer y such that $ny = au - 1 \Rightarrow n(-y) + au = 1 \Rightarrow bn(-y) + bau = b \Rightarrow a \cdot bu - b = bny \Rightarrow b \cdot u$ is solution of the congruence.

Remark :
We will note $u = a^{-1}$ thereafter. Hence $x = ba^{-1}$ is solution of the congruence $ax \equiv b \ (\mod n)$.

Example :

We want to solve $9x \equiv 15$ (mod 28).

We can show that this congruence is equivalent to $3x \equiv 5$ (mod 28) (exercice). The inverse of 3 modulo 28 is 19 (exercice), then, $x = 5 \cdot 19 = 95$ is one solution. There are therefore an infinite number of solutions, all congruent to **11** modulo 28 !

4.3.2 Chinese Remainder Theorem

Now that we know how to solve a linear congruence, let us examine solving a system of linear congruences.

Example :

$$\left[\begin{array}{l} x \equiv 5 (\quad \mathrm{mod}\ 11) \\ x \equiv 3 (\quad \mathrm{mod}\ 10) \end{array} \right. \tag{4.6}$$

One possibility is to look for multiples of 11 plus 5 and stop at the first number ending in 3.

16; 27; 38; 49; 60; 71; 82; **93**... Thus, there are an infinite number of solutions congruent to 93 modulo 110.

But this method is not generalized to more difficult systems or with more equations !

Then we have the **Chinese Remainder Theorem**

Let $m_1, m_2, ..., m_n$ positive integers, pairwise coprime, and $b_1, b_2, ..., b_n$ arbitrary integers. Then the linear system of simultaneous congruences

$$\left[\begin{array}{l} x \equiv b_1 (\quad \mathrm{mod}\ m_1) \\ x \equiv b_2 (\quad \mathrm{mod}\ m_2) \\ \qquad\qquad ... \\ x \equiv b_n (\quad \mathrm{mod}\ m_n) \end{array} \right. \tag{4.7}$$

has a unique solution modulo $M = m_1 \cdot m_2 \cdot ... \cdot m_n$

We set $M_i = \dfrac{M}{m_i}$ for each i with $1 \le i \le n$. M_i is an integer by definition of M.

Since m_i are pairwise coprime, we have $GCD(M_i, m_i) = 1$, then, each M_i has an inverse x_i modulo m_i.

We define last $x = b_1 \cdot M_1 \cdot x_1 + b_2 \cdot M_2 \cdot x_2 + ... + b_n \cdot M_n \cdot x_n$.

Since $M_i \cdot x_i = 1$ (mod m_i) for each i with $1 \le i \le n$ and m_i divides M_j for each $i \ne j$, x is a solution of the n congruences.

To demonstrate the *uniqueness* modulo M, we need the following result :

Lemma

> If $a \equiv b \ (\ \mod r\ )$, $a \equiv b \ (\ \mod s\ )$ and $GCD(r,s) = 1$, then $a \equiv b \ (\ \mod r \cdot s\ )$

Demonstration :

There exists integers u and v such that $a - b = ur$ and $a - b = vs$, therefore $ur = vs$. As r and s are coprime, r divides v (or s divides u). Then there exists integer w such that $rw = v$. Finally, $a - b = vs = rws = wrs$, hence $a \equiv b \ (\ \mod r \cdot s\ )$.

Remark
This result allows us to say that if a number is divisible by r and s, then it is divisible by $r \cdot s$ provided $GCD(r,s) = 1$.

Example : an integer is divisible by 6 if it is divisible by 2 and by 3, because $GCD(2,3) = 1$. On the other hand, we cannot say that an integer is divisible by 18 if it is divisible by 3 and by 6 (24 is not divisible by 18).

To complete the proof of the **uniqueness of the Chinese Remainder Theorem** (4.7), suppose there are two solutions x and x', then $x \equiv x' \ (\ \mod m_i\ )$ for each i with $1 \leq i \leq n$.

By lemma above, $x \equiv x' \ (\ \mod m_1 \cdot m_2 \cdot .... \cdot m_n\ )$, then $x \equiv x' \ (\ \mod M\ )$.

Finally, consider the example (4.6), we have $M = 110$, $M_1 = 10$, $M_2 = 11$, $x_1 = 10$ because $10 \cdot 10 \equiv 1 \ (\ \mod 11\ )$, $x_2 = 11$ because $11 \cdot 11 \equiv 1 \ (\ \mod 10\ )$. Then $x = 5 \cdot 10 \cdot 10 + 3 \cdot 11 \cdot 11 = 863$. Now, $863 \equiv 93 \ (\ \mod 11 \cdot 10\ )$.
To check the calculations (inverse, modulo), you can use [ ] or [ ]

4.4   Euler's totient function

We saw that two number a and b are **coprime** when $GCD(a,b) = 1$. Given an interger n, it can be useful to know how many non-zero integers not bigger than n are coprime to n.

Examples :
- $n = 8$: coprime integers to 8 are 1, 3, 5, 7 (there are 4 integers coprime to 8)
- $n = 20$: coprime integers to 20 are 1, 3, 7, 9, 11, 13, 17 and 19 (8 integers coprime to 20)

We already saw that $GCD(r,n) = 1 \Leftrightarrow r$ is a unit of $\mathbb{Z}/_n\mathbb{Z}$. Then, the number of non-zero integers not bigger than n and coprime to n is the **number of units** of $\mathbb{Z}/_n\mathbb{Z}$.

Definition of the Euler's totient function

The *Euler's totient function* is the function that associate to every non-zero integer n the number of units of $\mathbb{Z}/_n\mathbb{Z}$.

This function is written $\phi(n)$.

Here is a graph of this function for $2 \leq n \leq 100$:

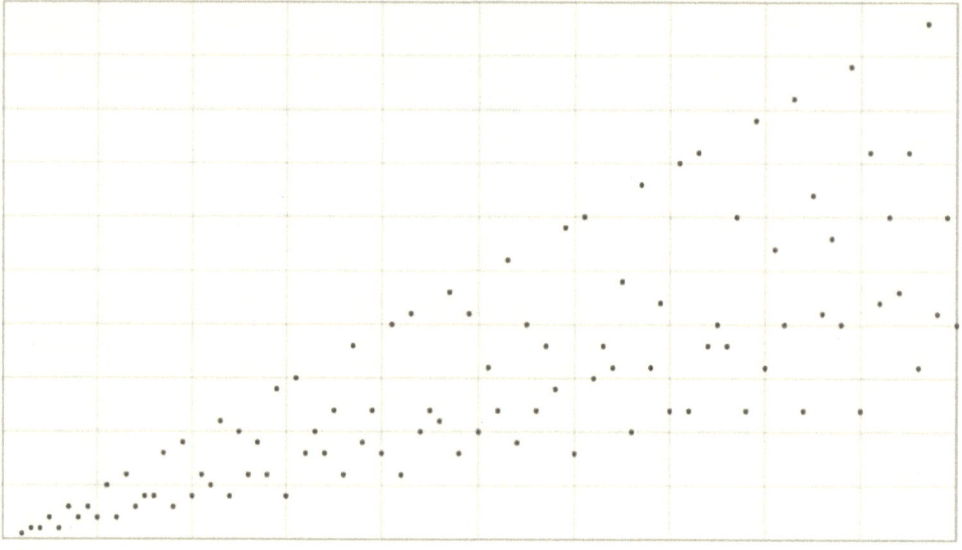

Isn't it a wonderful function ?! (widget in [2])

The dots at the top correspond to the value for prime numbers.

Properties

$$\phi(p) = p - 1 \text{ , if } p \text{ is prime}$$

$$\phi(p^r) = p^{r-1}(p-1) = p^r\left(1 - \frac{1}{p}\right) \text{ , if } p \text{ is prime and } r \in \mathbb{N}$$

$$\phi(m \cdot n) = \phi(m) \cdot \phi(n) \text{ , if } GCD(m,n) = 1 \text{ and } m,n \in \mathbb{N}$$

For the proof, you can see [1].

Examples :

- $n = 8$: $\phi(8) = \phi\left(2^3\right) = 2^2 \cdot (2-1) = 4$
- $n = 20$: $\phi(20) = \phi\left(2^2 \cdot 5\right) = \phi\left(2^2\right) \cdot \phi(5) = 2 \cdot 4 = 8$
- $n = 1000$: $\phi(1000) = \phi\left(2^3 \cdot 5^3\right) = \phi\left(2^3\right) \cdot \phi\left(5^3\right) = 4 \cdot 100 = 400$

There is a formula to symplify the calculations for large numbers :

If $n = p_1^{a_1} \cdot p_2^{a_2} \cdot .... \cdot p_r^{a_r}$, where p_i are all prime and distinct. Then

$$\phi(n) = n \cdot \left(1 - \frac{1}{p_1}\right) \cdot \left(1 - \frac{1}{p_2}\right) \cdot .... \cdot \left(1 - \frac{1}{p_r}\right)$$

This way, we don't need to consider exponents !

Examples :

- $n = 1000$: $\phi(1000) = \phi\left(2^3 \cdot 5^3\right) = 1000 \cdot \left(1 - \frac{1}{2}\right) \cdot \left(1 - \frac{1}{5}\right) = 1000 \cdot \frac{1}{2} \cdot \frac{4}{5} = 400$

- $n = 3600$: $\phi(3600) = \phi\left(2^4 \cdot 3^2 \cdot 5^2\right) = 3600 \cdot \frac{1}{2} \cdot \frac{2}{3} \cdot \frac{4}{5} = 960$

- $p = 1013$: $\phi(1013) = 1012$, because 1013 is prime.

4.4.1   Euler's Theorem

Given $a \in \mathbb{Z}$ and $n > 1$,

$$\text{If } GCD(a,n) = 1, \text{ then } a^{\phi(n)} \equiv 1 \ (\mod n)$$

You cas see [ ] for the demonstration.

Corollary

$$\text{If } GCD(a,n) = 1, \text{ then } a^{-1} \equiv a^{\phi(n)-1} \ (\mod n)$$

Demonstration :
$a^{\phi(n)} \equiv 1 \ (\mod n) \Rightarrow a \cdot a^{\phi(n)-1} \equiv 1 \ (\mod n)$, hence the result.

Corollary
If $GCD(a,n) = 1$, then the solution of the linear congruence $ax \equiv b \ (\mod n)$ is

$$x \equiv b \cdot a^{\phi(n)-1} \ (\mod n)$$

Example :
Solve the congruence $18x \equiv 15 \ (\mod 13)$.
We have $x \equiv 15 \cdot 18^{12-1} \ (\mod 13) \equiv 2 \cdot 5^{11} \ (\mod 13) \equiv 2 \cdot (5^2)^5 \cdot 5 \ (\mod 13)$
$\equiv 2 \cdot (-1)^5 \cdot 5 \ (\mod 13) \equiv 2 \cdot (-5) \ (\mod 13) \equiv (-10) \ (\mod 13) \equiv 3 \ (\mod 13)$

4.4.2   Fermat's Little Theorem

Given p prime and a non-zero integer,

$$\text{If } p \text{ not divides } a \text{, then } a^{p-1} \equiv 1 \ (\ \text{mod } p \ )$$

This theorem result from Euler's Theorem.

Remark :
This theorem is equivalent to the following variant :

Given p prime and a any integer, then

$$a^p \equiv a \ (\ \text{mod } p \ )$$

In particular, if a is invertible modulo p, we have back Fermat's little theorem by multiplying the equation by a^{-1}. If p divides a, then $a \equiv 0 \ (\ \text{mod } p \ )$ and the congruence is also true.

Examples :
- $14^3 \equiv 14 \ (\ \text{mod } 3 \ ) \equiv 2 \ (\ \text{mod } 3 \ )$
- $2^{97} \equiv 2 \ (\ \text{mod } 97 \ ) \Rightarrow 2^{97} - 2$ is divisible by 97
- $16^{77} \ (\ \text{mod } 7 \ ) \equiv 2^{7 \cdot 11} \ (\ \text{mod } 7 \ ) \equiv 2^{11} \ (\ \text{mod } 7 \ ) \equiv 2^7 \cdot 2^4 \ (\ \text{mod } 7 \ ) \equiv 2^5 \ (\ \text{mod } 7 \ ) \equiv 4 \ (\ \text{mod } 7 \ )$

4.4.3   Primality Test and Binary Exponentiation

Given $a = 2$ and $n > 2$, we get a primality test. If $2^n \not\equiv 2 \ (\ \text{mod } n \ )$ by contrapositive of Fermat's little theorem, n is not prime.

Example :
Is $n = 1027$ prime ?
We have to calculate $2^{1027} \ (\ \text{mod } 1027 \ )$ and see if the result is equal to $2 \ (\ \text{mod } 1027 \ )$:

Without a modular calculator [ ], we can do what is called **binary exponentiation**.

We begin to decompose 1027 into a sum of powers of 2.

We get $1027 = 1024 + 2 + 1 = 2^{10} + 2^1 + 2^0$, so $2^{1027} = 2^{2^{10}} \cdot 2^{2^1} \cdot 2^{2^0}$.
In fact, we can write a table and, by sqaring a line to another, we can make calculus with a traditional calculator :

quotient	rest	m	2^{2^m} (mod 1027)
1027	1	0	2
513	1	1	4
256	0	2	16
128	0	3	256
64	0	4	835
32	0	5	919
16	0	6	367
8	0	7	152
4	0	8	510
2	0	9	269
1	1	10	471

Then, we have to calculate :

$$2^{1027} \ (\ \text{mod } 1027\ ) \equiv 2^{2^{10}} \cdot 2^{2^1} \cdot 2^{2^0} \ (\ \text{mod } 1027\ ) \equiv 471 \cdot 4 \cdot 2 \ (\ \text{mod } 1027\ ) \equiv 687 \ (\ \text{mod } 1027\ )$$

Thus 1027 is not prime.

4.5   RSA Theorem - Rivest Shamir Adleman

Given two distinct primes p and q, let $n = p \cdot q$.

If e is coprime to $\phi(n) = (p-1)(q-1)$ and d is such that $e \cdot d \equiv 1 \ (\ \text{mod } \phi(n))$,
then, for all integer m,

$$m^{ed} \equiv m \ (\ \text{mod } n\ )$$

Remark : n and e constitutes the **Public Key** and p, q and d the **Private Key**. The public key can be published (e.g. on the internet), but the private key must be kept secret.

Example : $n = 2777 \cdot 3331 = 9,250,187$ and $\phi(n) = (p-1)(q-1) = 2776 \cdot 3330 = 9,244,080$, we take, by example, $e = 247$ and we verify that $GCD(e, \phi(n)) = 1$ with the Euclid's algorithm. We can calculate then d with the **extended Euclidean algorithm** (4.2.8), since there exists d and y such that $d \cdot e + y \cdot \phi(n) = 1 \ \Rightarrow \ d = 7,747,063$.

Suppose $m = 12345$, then $m' \equiv m^{247} \ (\ \text{mod } n\ ) \equiv 1,974,687 \ (\ \text{mod } n\ )$ is the encrypted message.

Then, $1,974,687^{7,747,063} \ (\ \text{mod } n\ ) \equiv 12345 \ (\ \text{mod } n\ )$ and we find again the decrypted message m.

To do these calculations, you can use an *computer algebra system* **CAS** (like Maple or WolframAlpha) or the RSA calculator widget of [1], or the iPad App [4]. These calculations are in fact infeasible with a traditional calculator.

It is important to understand that the safety of RSA is based on the difficulty of factoring the number n and thus finding p and q that allow us to calculate d. However, the above RSA key contains only 7 digits and most CAS allow you to find its factorization in a few thousandths of a second.

4.5.1   Public Key and Certification Authorities

Most RSA keys in communications today are based on 2048-bit keys, i.e., keys equivalent to 617 digits in base 10.

Current computers, including supercomputers, are not capable of factoring such numbers in a reasonable amount of time (it would take years to do so).

It is therefore necessary for RSA keys to be changed regularly and they are managed by *certification authorities*.

Soon, 2048-bit keys will no longer be sufficient and will give way to 4096-bit keys. But all this is likely to change with the advent of consumer quantum computers that will be able, equipped with the right algorithms, to reduce computing times sufficiently so that the security of a 4096-bit key or more can no longer be guaranteed.

To see the public key of a bank, for example, just click on the small padlock that appears in the url and view the details.

This is an RSA key of 2048 bits or 256 hexadecimal bytes. The exponent e is 65,537. The certification authorities is DigiCert. The expiry date of the certificate can also be found above.

4.5.2 RSA Encryption Process

Here is an illustration of these steps in practice, assuming that a sender A wants to send a message by email to a recipient B who has therefore published his public key on the internet.

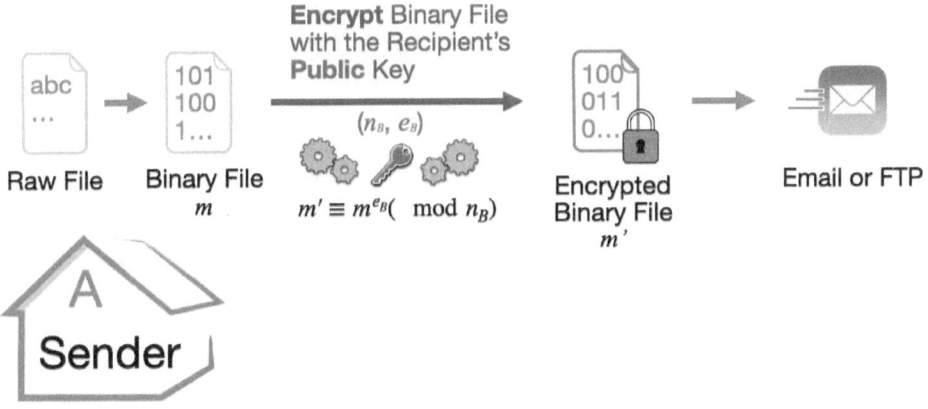

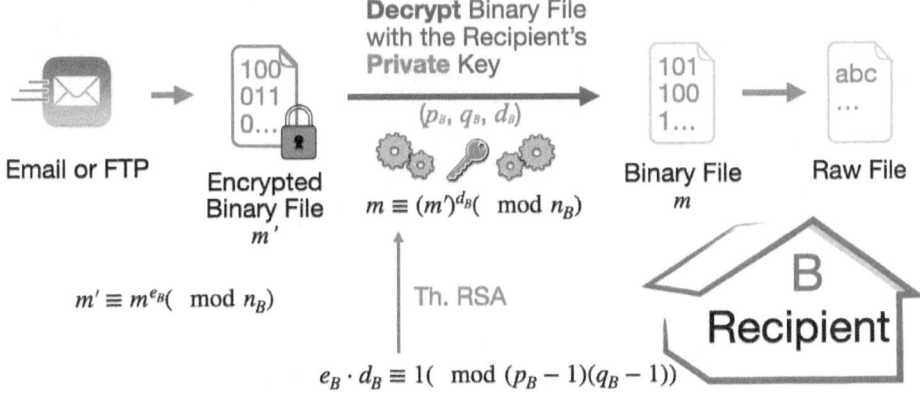

4.5.3   RSA Signature Process

The principle of signing a message allows the receiver to ensure that the message comes from the right person. This is called **non-repudiation** of data.

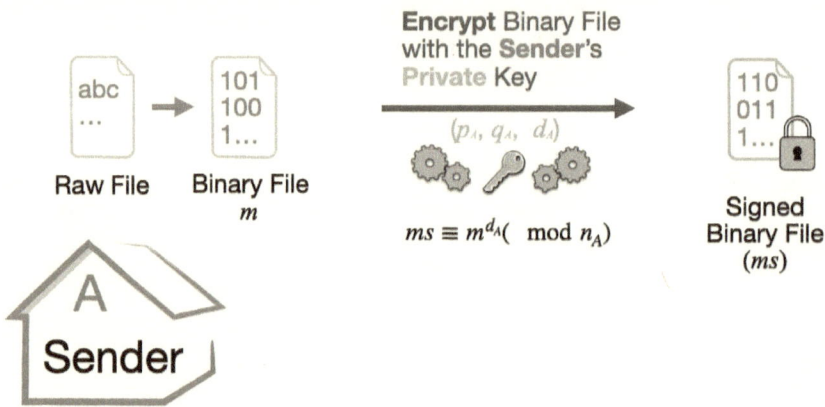

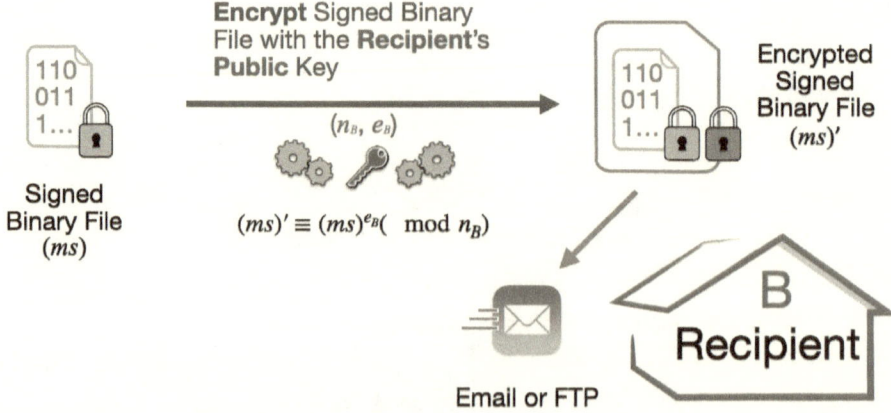

It is necessary to do these two steps in that order (first signing), otherwise any malicious person intercepting the message could change the signature with the sender's public key !

THE RSA SIGNATURE PROCESS 3

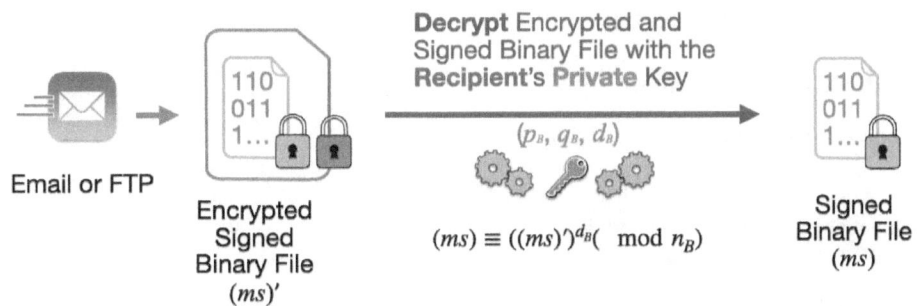

Email or FTP

Encrypted
Signed
Binary File
$(ms)'$

Decrypt Encrypted and
Signed Binary File with the
Recipient's Private Key

(p_B, q_B, d_B)

$$(ms) \equiv ((ms)')^{d_B}(\mod n_B)$$

Signed
Binary File
(ms)

THE RSA SIGNATURE PROCESS 4

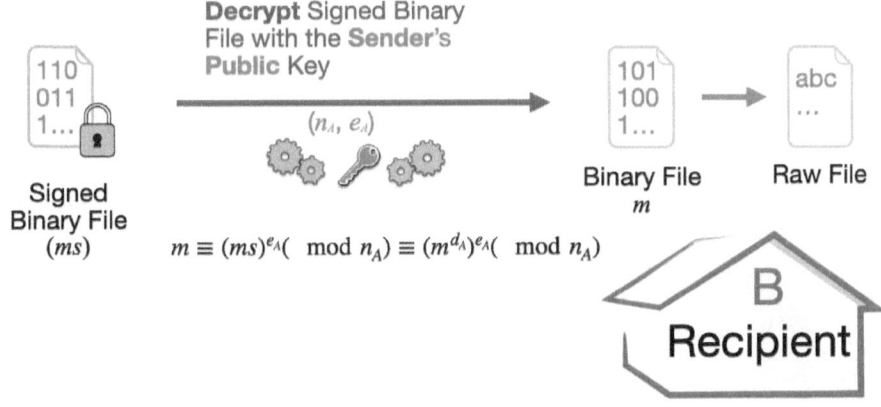

Signed
Binary File
(ms)

Decrypt Signed Binary
File with the **Sender's**
Public Key

(n_A, e_A)

Binary File
m

Raw File

$$m \equiv (ms)^{e_A}(\mod n_A) \equiv (m^{d_A})^{e_A}(\mod n_A)$$

B
Recipient

If you want to sign and encrypt an email yourself before sending it, this is possible with **OpenPGP** (openpgp.org/software).

In practice, for example when you make online payments on your bank's website, data security is more user-friendly for the user. However, there are 4 main steps : **authentication** (sms, PIN code, QR code, etc.), **integrity** of the data sent (fingerprint unique to each message which makes possible to check that the content of the message has not been modified during shipment), data **encryption** and **non-repudiation** (4.5.2 and 4.5.3 above).

In fact, RSA is not used to encrypt all data exchanged because it would take too much computing time, but only to exchange symmetric keys (SHA-256) between interfaces. Then, calculation algorithms (hash functions) take over to encrypt/decrypt the exchanged data much faster.

4.5.4   Important Remarks

If $m = 0 \ (\mod p)$ or $m = 0 \ (\mod q)$, the RSA theorem still works.

In practice, m must be smaller than n, because if not, it's representative modulo n who will appear after decrypting may not correspond to the original message m. That said, with a 2048-bit key, the possible length of the message is already very large.

e and d are units of $\mathbb{Z}/_{\phi(n)}\mathbb{Z}$, there are $\phi(\phi(n))$ possible choices for e, but you better not take an integer too small or too large, to also save the time of computers. In general, $e = 65,537 = 2^{16} + 1$ (prime number).

Index

Bibliography

[1] Pascal Cardin. *Number Theory & RSA Cryptography*. iBooks Store Apple, 2018.

[2] Pascal Cardin. *Number Theory & RSA Cryptography Student Version*. iBooks Store Apple, 2018.

[3] Alain Cardinaux. *CalcMod-Light*. Apple iPhone iPad, 2018.

[4] Alain Cardinaux. *CalcMod-RSA*. Apple iPad, 2018.